Natural Stone and World Heritage: Delhi-Agra, India

Natural Stone and World Heritage

Edited by Dolores Pereira

Department of Geology, University of Salamanca, Salamanca, Spain

ISSN: 2640-0162
eISSN: 2640-0170

Volume 2

Natural Stone and World Heritage

Delhi-Agra, India

Gurmeet Kaur, Sakoon Singh,
Anuvinder Ahuja and
Noor Dasmesh Singh

CRC Press is an imprint of the
Taylor & Francis Group, an **informa** business
A BALKEMA BOOK

Cover image: courtesy of Gurnoor Arora (NAFA Singapore)

CRC Press/Balkema is an imprint of the Taylor & Francis Group, an informa business

Typeset by Apex CoVantage, LLC

Library of Congress Cataloging-in-Publication Data

Applied for

Published by: CRC Press/Balkema
Schipholweg 107C, 2316XC Leiden, The Netherlands
e-mail: Pub.NL@taylorandfrancis.com
www.crcpress.com – www.taylorandfrancis.com

ISBN: 978-0-367-25180-2 (Hbk)
ISBN: 978-0-429-28639-1 (eBook)

DOI: 10.1201/9780429286391
https://doi.org/10.1201/9780429286391

To Our Parents

Contents

	Foreword	ix
	Preface	x
	Acknowledgements	xiii
	List of acronyms	xv
1	UNESCO World Heritage Sites, International Union of Geological Sciences and Heritage Stone Subcommission	1
2	Delhi and Agra vis-à-vis monuments	23
3	Repository of stones used in Delhi and Agra UNESCO Sites: Aravalli Mountain Belt and Vindhyan Basin	37
4	UNESCO Heritage Sites of Delhi and Agra: An account	60
5	Historical quarries of the Makrana Marble, the Vindhyan Sandstone and the Delhi Quartzite: An account	109
6	Preservation, conservation and restoration of UNESCO World Heritage Sites of Delhi and Agra	142
7	Conclusions	151

References 156
Glossary 168
Natural Stone and World Heritage 171

Foreword

The famous monuments and buildings in India, such as Taj Mahal, Qutb Minar, the Humayun's Tomb and the Victoria memorial in Kolkata, are widely known and recognized as Indian heritage. However, the construction of such historic buildings, the stone that was used for their construction and the meaning of the stone through its use over the centuries are details that, at least for the vast majority, are unknown. Now this book brings us closer to the historical constructions of emblematic places such as Delhi and Agra, with the UNESCO recognized monuments Humayun's Tomb, made of red sandstone, and the Taj Mahal, made of Makrana marble. But many others will see light that without a publication like this would continue to have an unfairly limited diffusion. This book is another step in the recognition of natural stone and the heritage built on stone.

Prof. Dolores Pereira
Secretary General, Heritage Stone Subcommission
Chair, IUGS Publications Committee

Preface

Though there is widespread interest in the heritage monuments, there is almost total ignorance regarding the construction material used therein. Thus, a series of books has been proposed to be published on Natural Stone and World Heritage by CRC Press/Balkema, Taylor and Francis Group. As a follow-up *Natural Stone and World Heritage: Salamanca (Spain)* by Prof. Dolores Pereira became the first publication in the year 2019.

The present book, second in the series, aims to create awareness on the stone-built World Heritage Sites in Delhi and Agra. Numerous books on historical accounts and architectural details of the World Heritage Sites of Delhi and Agra are available, but they lack details on the building stones and historical accounts of quarries. This book documents the building stone repositories, which contributed material for construction of splendid architectural heritages. The aspect of blending architecture and the choice of stone based on various criteria have been highlighted.

An attempt has been made to integrate the natural stones and architectural heritage by following a holistic approach, which hopefully will cater to the curiosity of readers having diverse backgrounds. It covers a wide spectrum of World Heritage Sites constructed for religious, funerary, military and residential needs during the Sultanate and Mughal periods. All the significant monuments valuable for their architectural and historical existence have been discussed.

The present book includes seven chapters giving a comprehensive account of natural stone used in construction of the World Heritage Sites in Delhi and Agra. Chapter 1 discusses the role of the UNESCO in designating and preserving the World Heritage Sites and of the Heritage Stone Subcommission (HSS) of International Union of Geological Sciences (IUGS) in bringing to the fore the heritage stones used in the architectural heritage built in the past with a cultural link. The definition of the term Global Heritage Stone Resource (GHSR) and Terms of Reference for designation of natural stones as GHSRs are touched upon. A concise summary of the 22 designated GHSR from around the world is presented. Numerous Indian examples are listed that merit GHSR designation. Chapter 2 deals with a significant historical and cultural account of these two cities to give the reader a continuum in terms of space and time that led to the stone-built architectural heritage. The account on "nine cities of Delhi" followed by the city of Agra is discussed to engage the readers to give them an overview of the two historical cities. Chapter 3 gives a comprehensive account on the geology of the natural stones vis-à-vis the geological and geographical account of the cities of Delhi and Agra. The Aravalli Mountain Belt and the Vindhyan Basin repositories of the natural stones used in the World Heritage Sites of Delhi and Agra also find a place in this chapter. This chapter gives a better understanding of building stones used in the World Heritage Sites of these two cities, accompanied by maps, field photographs and petrographic details of these stones. Chapter 4 incorporates focussed accounts of the Qutb Minar and its adjoining monuments, the Humayun's Tomb and the Red Fort Complex, in Delhi; and the Agra Fort and the Taj Mahal in Agra. Each description includes details on the main monument and the adjunct complexes. It attempts to contextualize each monument within a matrix of geological, geographical, historical, cultural and architectural influences that marked its construction and evolution through time. Detailed plans and photographs pertaining to each monument are provided for an all-inclusive understanding. Chapter 5 elaborates the historical and present status of the quarries and their functionality vis-à-vis the UNESCO World Heritage Sites of Delhi and Agra. Description of the Makrana marble and the Delhi quartzite of Aravalli Mountain Belt and the red sandstone of the Vindhyan Basin are presented with excerpts from the narratives of 16th and 17th century native and foreign scholars. Illustrative maps of quarry sites from the Mughal and Colonial periods adorn this chapter, which is the first attempt of this type. Numerous subordinate stones used in these monuments also find a mention. Chapter 6 touches upon various charters framed for the upkeep of the Heritage Sites of India, roles of Archaeological Survey of India and organizations like INTACH dealing with Heritage monuments of India. Conservation, preservation and restoration measures undertaken for the World Heritage Sites of Delhi and Agra are provided with illustrations. Chapter 7 integrates various aspects of the natural stones, history and architecture of the World Heritage Sites of Delhi and Agra. It establishes the significance of natural stones vis-à-vis the architectural heritage in space and time in these two historical cities.

The book is the first of its kind where building stones used in the construction of the World Heritage Sites of Delhi and Agra are enumerated with a focus on the historical quarries. It adds a new dimension to the study of building stones used in the World Heritage Sites with reference to the cultural, historical and architectural accounts of these monuments.

Acknowledgements

Gurmeet Kaur (GK) would have never dreamt of writing a book on natural stones and UNESCO World Heritage Sites of Delhi and Agra, but for the prompting by Prof. Dolores Pereira, to whom she is deeply beholden. Her constant encouragement and invaluable guidance helped in accomplishing this arduous task. GK is thankful to her co-authors Sakoon Singh (SS), Anuvinder Ahuja (AA) and Noor Dasmesh Singh (NDS) for their relevant contributions to this book. GK is highly obliged to Dr O.N. Bhargava (Emeritus Professor, Department of Geology, Panjab University, Chandigarh) for his help in reviewing Chapters 3 and 5 and his overall guidance. She is most thankful to Ms Gurnoor Arora from NAFA (Singapore) for contributing the attractive cover page of the book.

GK is beholden to the honorable Vice-Chancellor (Prof. Raj Kumar), Panjab University and Chairperson (Prof. Rajeev Patnaik), Department of Geology, Panjab University, for granting permission to carry out field visits to the heritage monuments and quarry sites relevant to her project on heritage stones. GK is grateful to Dr Usha Sharma (IAS), Director General, Archaeological Survey of India (ASI) and other ASI officials (Shri Janhwij Sharma, Shri T.J. Alone, Dr N.K. Pathak, Shri Daljit Singh, Shri Manuel Joseph and Shri Manoj Kumar) including the chief librarian (Smt Savita Kaul) for their generous support in collection of information pertaining to the UNESCO monuments of Delhi and Agra. AA wishes to acknowledge staff of the American Embassy School (AES) library for generously providing books for reference during the course of writing this book. GK extends her sincere thanks to Mr Mukul Rastogi (CEO, CDOS), Mr Kireet Acharya and Mr Pradeep Agrawal (Acting Director CDOS, Jaipur) for helping her with data on physico-mechanical properties of the stones from Makrana and western Vindhyan Basin. GK expresses her gratitude to several other colleagues and students with whom she has been collaborating in dealing with heritage stones of India.

GK extends special thanks to her four research scholars: Ms Parminder Kaur, Ms Jaspreet Saini, Mr Amritpaul Singh and Mr Sanchit Garg for drafting most of the maps and plans used in this book. The unstinting help of Ms Parminder Kaur and Ms Jaspreet Saini with accumulation of relevant literature from the libraries of the ASI at Delhi and Agra is gratefully acknowledged. GK and AA are grateful to Mr Amarjit Singh Ahuja for his help during the visits to former quartzite quarry sites of Lal Kuan and heritage monuments of Delhi. GK and AA benefitted from an initial meeting with Mr Vikramjeet Singh Rooprai whom they wish to acknowledge. Mr Rajeev, Department of Geology, Panjab University, is thanked for preparing thin sections of rocks for petrographic studies.

GK expresses gratitude to her parents, Mrs Satnam Kaur and Mr D.S. Arora, and sister Satwinder Kaur for their affectionate care and encouragement, which sustained her throughout this work.

SS, AA and NDS express their gratitude, foremost to Dr Gurmeet Kaur, Assistant Professor at the Department of Geology, Panjab University, Chandigarh, for initiating this wonderful project that brought together professionals from diverse backgrounds. The most exciting part of this exploration has been a deep engagement, with a multi-disciplinary ethos that has provided a dynamic learning curve to all of us. The intensity of engagement with the UNESCO World Heritage of Delhi and Agra has been deeply illuminating and will stay as a gift to cherish for a lifetime. No words would suffice to express the gratitude to our parents and families for their support and encouragement rendered during the course of writing this book.

List of acronyms

AMB	Aravalli Mountain Belt
ASI	Archaeological Survey of India
BGC	Banded Gneissic Complex
DFB	Delhi Fold Belt
GHSR	Global Heritage Stone Resource
HSCS	Heritage Sites and Collections Subcommission
HSS	Heritage Stone Subcommission
ICOMOS	International Council on Monuments and Sites
ICG	International Commission on Geoheritage
INTACH	Indian National Trust for Art and Cultural Heritage
IUGS	International Union of Geological Sciences
NCT	National Capital Territory
NDFB	Northern Delhi Fold Belt
SDFB	Southern Delhi Fold Belt
UNESCO	United Nations Educational, Scientific and Cultural Organization

Chapter 1

UNESCO World Heritage Sites, International Union of Geological Sciences and Heritage Stone Subcommission

> "*We talk with reason when we say that nothing is more clearly written than what is written in stone.*"
>
> Gustave le Bon

1.1 Introduction

The natural stones have been an integral part of human civilizations. Rocks and stones were put to use by prehistoric man in the form of cave dwellings, tools for hunting and fire for cooking and keeping warm. Subsequently cavemen progressed from nomadic life patterns to settled patterns, promoting community dwellings with occupations such as rearing of animals, agriculture, manufacturing of artifacts from minerals, rocks and their derivatives for multiple uses. The natural stones and rocks have played a pivotal role in the evolution of the civilized society from antiquity to the present day. The different civilizations around the globe record usage of stone in the construction of forts, palaces, temples, mosques, cathedrals, synagogues, citadels, tombs, mausoleums, dwellings etc. reflecting their intrinsic cultures and traditions. However, the common attributes that link the heritage stones globally are their aesthetics and architectural creativity.

Disparate civilizations all over the world have left a rich legacy of stone-built architectural heritage and shed light on art and design, tradition, religious value and splendor. Rock-cut architecture has been unearthed across civilizations and has existed as a mode of expression in the past primarily for promoting religion, social demeanor, good ethics and awareness (Kaur *et al.*, 2019a; Kaur *et al.*, 2019b). Huge structures and sculptures were created by skilled craftsmen to depict the inimitable way of life of their times. Many architectural structures were erected and engraved with figures from time to time with rock as a medium of expression. The artisans and sculptors experimented with a variety of stones available in close proximity. If we look at the chronological order of the ancient architecture we can understand the innovative approach and significance of art and architecture from times beyond.

1.2 United Nations Educational, Scientific and Cultural Organization (UNESCO)

UNESCO was founded on 16 November 1945 and the constitution of UNESCO, signed on 16 November 1945, came into force on 4 November 1946.

UNESCO World Heritage Committee comprises 21 state parties, elected by their General Assembly. The task of this committee is to unanimously recognize and take responsibility for saving select samples of art and architecture for the benefit of humanity. The criteria for a site to be a suitable candidate for World Heritage Site is that it needs to have unique characteristics imbued with historical, geographical, ecological and literary significance and have been an embodiment of spectacular work for a multitude of generations. Any piece of architecture, nature or literary work that proves to be an evidence of history and indicates an evolution of different art forms, or was used as an expression to communicate religious, social and cultural values in ancient times, which needs protection for longevity and needs to be preserved for the generations to come are protected by UNESCO. It takes care of such heritage landmarks and protects them from human negligence and ignorance, weathering due to climatic changes, natural hazards and pollution (https://whc.unesco.org/en/convention/).

UNESCO is working its way towards recognizing and protecting more and more Heritage Sites of cultural, natural and mixed importance from different countries, and the list is getting bigger each year. All the recognized monuments embody a wealth of creativity, innovation, perspective and intellect and symbolize a larger value which is an inspiration to the modern day architects, sculptors and artisans. There are 167 state parties with inscribed World Heritage Sites. So far the total World Heritage Sites recognized by UNESCO is 1121 which includes 213 Natural Sites, 869 Cultural Sites and 39 Mixed Sites from all regions (Fig. 1.1; Table 1.1; https://whc.unesco.org/en/list/).

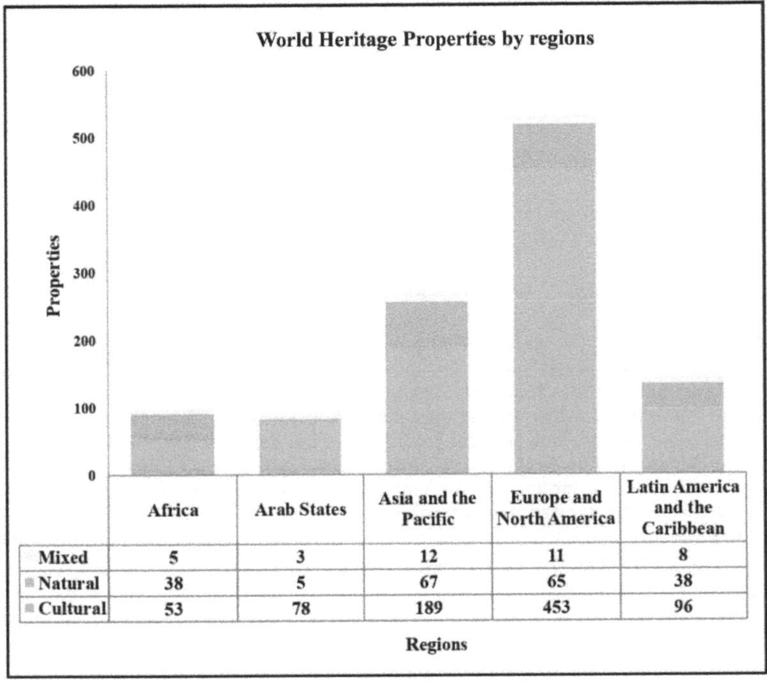

Figure 1.1 UNESCO World Heritage Sites by region in 2018
(Source: https://whc.unesco.org/en/list/stat, accessed September 30 2019)

The present book on Natural Stone and World Heritage of Delhi and Agra is part of a series of books to be published on Natural Stone and World Heritage by CRC Press/Balkema, Taylor and Francis Group. The first book *Natural Stone and World Heritage: Salamanca (Spain)* focuses on vital aspects of the natural stones used in the built architectural heritage of Salamanca (Pereira, 2019). The current book is along the lines of the first book where a comprehensive account of natural stone and its built World Heritage Sites are dealt with. The foremost inventiveness of this book series is to promote the natural stones used in the World Heritage Sites.

Table 1.1 World Heritage Properties by region

	Cultural	*Natural*	*Mixed*	*Total*	*%*	*State parties with inscribed properties*
Africa	53	38	5	96	8.56%	35
Arab States	78	5	3	86	7.67%	18
Asia and the Pacific	189	67	12	268*	23.91%	36
Europe and North America	453	65	11	529*	47.19%	50
Latin America and the Caribbean	96	38	8	142*	12.67%	28
Total	869	213	39	1121	100%	167

* The properties "Uvs Nuur Basin" and "Landscapes of Dauria" (Mongolia, Russian Federation) are trans-regional properties located in Europe and Asia and the Pacific region. They are counted here in the Asia and the Pacific region.

* The property "The Architectural Work of Le Corbusier, an Outstanding Contribution to the Modern Movement" (Argentina, Belgium, France, Germany, India, Japan, Switzerland) is a trans-regional property located in Europe, Asia and the Pacific and Latin America and the Caribbean region. It is counted here in the Europe and North America.

(Source: https://whc.unesco.org/en/list/stat, accessed September 30 2019)

1.3 Stones: symbolic of architectural heritage

Rocks were cut and hewn with specialized tools to create monolithic and other structures. Stone masonry or stonecraft surfaced from different parts of the world which included the larger than life rock structures such as forts, citadels, palaces, churches, temples, monasteries, tombs and buildings of cultural, social and funerary significance. The artisans were well aware of the existing rock types, their sustenance and durability and chiseled them to create an astounding body of work. Initially the caves or caverns were used to exhibit the expression of the artist, which was mainly religious in nature and promoted religion or depicted the daily chores performed by the people during those days. One could see the replication of design and patterns created on the walls and ceilings, akin to a commentary on the design strategy the artisans used. Some of the brilliant rock-cut architecture from different time periods is still considered exemplary for leaving behind a massive body of work which gives us a view of their lifestyles, traditions, cultures, values and beliefs. Lycian Tombs in Turkey, Petra in Jordan and cave temples in India and China show different techniques of sculpting used by the artisans in those days. Following are a few examples of stone built heritage from across the globe, recognized as UNESCO World Heritage Sites for their uniqueness and architectural beauty.

1.3.1 World heritage monuments: global scenario

Rock-cut sanctuaries in Turkey date back to 1250 BCE. They find a mention in the Christian Old Testament. Bogazkoy has the archaeological remains of Hittites which are to the northwest of Yozgat with the preserved temples, fortifications and royal residences with rich ornamentation of the Lion's Gate and the Royal Gate, the ensemble of limestone rock art at Yazilikaya (https://nypost.com/2019/06/26/ancientturkish-rock-carvings-that-have-baffled-scientists-could-be-a-calendar/; www.britannica.com/place/Bogazkoy).

The limestone rock cut Tombs of Persian Kings at Naqsh-e Rostam in Iran date back to the 4th and 5th century BCE (www.britannica.com/topic/Sasanian-dynasty; www.britannica.com/place/Persepolis#ref31169). It is a necropolis (large, designed cemetery with elaborate tomb monuments) with four large tombs cut into the limestone cliff face. These are decorated with figures of the king, small figures giving tribute are the soldiers and officials and this differentiation can be seen through variation in size.

Rock-cut tombs from the city of Lycia from the 4th century BCE are interesting monuments as they depict the magnificent relics on the side of cliffs. The British Museum also has a collection of Lycian artifacts like the Tomb of Payava, a Lycian aristocrat, Nereid sculptures (daughters of sea deities Nereus and Doris), Lycian coins etc. in its precincts. The Lycian civilization flourished only in the Mediterranean region of Turkey. These remains include the Parliament Building in the capital city Patara. The Lycian tombs made in light grey limestone, fortresses that look over the cities and sea, show the unique features of the ancient civilization (Wedekind *et al.*, 2017; www.britannica.com/place/Myra).

Petra in Jordan is also known as the Rose City because of the color of the sandstone from which it was built. The original name of Petra was Raqmu. It was devastated under Roman rule. Petra had many churches from the Byzantine era that were buried due to an earthquake. This place slipped into oblivion until the 19th century. The excavation in 1958 and then in 1993 revealed the monuments with relevance to the political, social and religious traditions followed by the inhabitants (www.britannica.com/place/Petra-ancient-city-Jordan).

Mogao Buddhist caves in Dunhuang, China date back to 400 CE. Desert in Xinjiang region surrounds the caves from the Tang dynasty. The sandstone caves near the town of Dunhuang, in Gansu province, depict workers, warriors and the landscape of Mount Wutai with details of mountains, rivers, cit ies, temples, roads and caravans (https://oxfordre.com/ religion/abstract/10.1093/acrefore/9780199340378.001.0001/acrefore9780199340378-e-173?rskey=OkuAqh&result=16; www.britannica.com/place/Mogao-Caves).

Churches in Cappadoccia, Turkey, are made of sedimentary rocks that were formed from the volcano deposits some 3 to 9 million years ago around the Miocene to Pliocene time period. People from the villages then used their ingenuity to convert them into their dwellings, churches and monasteries for worship. The Goreme Open Air Museum is very famous as it houses 30 carved rock churches and chapels, and they are embellished with some beautiful frescoes from the 9th to 11th century (www.britannica.com/place/Cappadocia).

The Colosseum, one of the wonders of the world and a heritage monument, is in Rome, Italy. The Flavian emperors had it constructed between 70 and 80 CE. It is also known as Flavian amphitheater with a massive construction that could accommodate thousands of Romans for a variety of shows like gladiator contests, mock sea shows, dramas etc. In the 12th century the Frangipani family occupied the Colosseum and converted it into a castle after putting a defensive wall around it. A variety of building materials were used to build the Colosseum like limestone, travertine (a sedimentary rock), marble, tuff and bricks (Richardson, 1992). Due to an earthquake huge devastation was caused to the south side of the structure in 1349. The materials from the collapsed wall were used to build other churches, palaces and hospitals in Rome.

Angkor Wat, a UNESCO World Heritage Site, is in Cambodia, South East Asia. It is a huge Buddhist temple built in the 12th century by Suryawarman II. The outer wall and the temple were made from slabs of sandstone and bricks and tiles made of laterite soil. The inside of the structure was mainly done in wood and other materials. The high sandstone walls were surrounded with a ditch filled with water to protect the temples and inhabitants from internal and foreign attacks. The unsupervised vegetation, war and natural disasters like earthquake caused havoc on the structure of the monument. Only the outer wall of the structure stayed. During the French rule here in the early 19th century, efforts for restoration were made and tourism was promoted. Countries like India, Germany and France also made efforts at restoring the site, and its craftsmanship is still lauded by tourists (www.history.com/topics/landmarks/angkor-wat; www.britannica.com/topic/Angkor-Wat).

Westminster Abbey is a collegiate church of Saint Peter, London. It is an architectural marvel amongst religious buildings. King Henry in 1245 CE got the construction of the church started. The church was made in Gothic style. The two western towers of the abbey were constructed with Portland stone between 1722 CE and 1745 CE. These towers were the last addition to the abbey. Walls and the flooring were done with Purbeck marble, and the tombstones present at the abbey were also done in a variety of marble and other stones, like carboniferous limestone, black Tournai and Welsh slate (Marker, 2015; Hughes *et al.*, 2016). Sir

Edwin Lutyens, who also has several other spectacular buildings to his credit, designed the entrance hall of the church. In the Statesmen's aisle the statues are made up of Carrara marble; the lantern floor is made up of Cararra white and Belgian black (www.britannica.com/topic/Westmin ster-Abbey).

Giza Necropolis, a UNESCO World Heritage Site, exists on the Giza plateau and dates from 2550 to 2490 BCE. The complex consists of the Great Pyramid of Giza, the Pyramid of Khafre, the Pyramid of Menkaure and the Great Sphinx of Giza (www.national geographic.com/archaeology-and-history/archaeology/giza-pyramids/). Limestone was used for the exterior of the pyramids; granite was used for burials; basalt was used for floors of temples and tombs; travertine or Egyptian alabaster, a type of limestone, was used for making sacred tools for mummification; and porphyry was used to make the high columns and wall linings.

1.3.2 World Heritage Sites: Indian scenario

UNESCO Heritage lists 38 Cultural, Natural and Mixed Sites in India (Fig. 1.2; https://whc.unesco.org/en/list/). Some Heritage Sites that continue to inspire us after centuries, generation after generation are discussed in the following, for their remarkable craftsmanship and architecture.

Rock shelters of Bhimbetka present on the southern edge of Vindhyan Range are in Central India. These cave paintings were based on themes, such as animals, early evidence of dance and hunting strategies and depiction of day to day life (Kaur *et al.*, 2019b).

Barabar caves are located in Bihar, and date back to 250-257 BCE (Smith, 1999). They are the oldest rock hewn caves that have Ashokan inscriptions preserved. These caves are present on the Barabar and Nagarjini hills. Some historians believe the caves at Barabar were excavated in granite and have an echo effect while Smith claims that they were excavated in hardcore quartzose gneiss. The Ashokan inscriptions were engraved on cave walls, boulders, columns and rock. Some caves also have Buddhist and Hindu sculptures, which were added in later years.

Ajanta, Ellora and Elephanta caves in Maharashtra are rock-cut structures in basalt of the Deccan plateau (Kaur *et al.*, 2019a). Ellora Caves include 34 rock-cut temples made in basalt. The Buddhist caves came up around 200 to 600 BCE while the Hindu temples came up around 500–900 CE with some sleeping cells specially constructed for travelling monks. The Kailasha temple dedicated to Lord Shiva in cave 16 is the most popular for its downward carving of monolithic basalt rock (www.britannica.com/place/Ellora-Caves). Ajanta is mainly known for its colorful Buddhist frescoes. There are 30 caves in Ajanta; some were constructed around 1st to 2nd century BCE while others got added

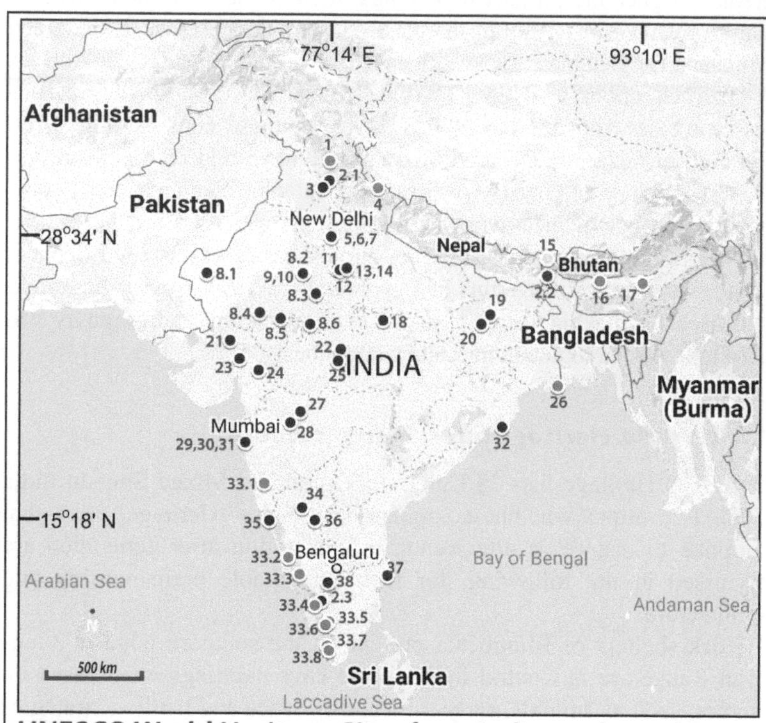

UNESCO World Heritage Sites from Indian subcontinent

1. Great Himalayan National Park Conservation Area 2. Mountain Railways of India 2.1 Kalka Shimla Railway, Himachal Pradesh 2.2 Daarjeeling Himalayan Railway, West Bengal 2.3 Nilgiri Mountain Railway, Tamilnadu 3. The Architectural Work of Le Corbusier, an Outstanding Contribution to Modern Movement 4. Nanda Devi and Valley of Flowers National Parks 5. Humayun's Tomb, Delhi 6. Qutb Minar and its monuments, Delhi 7. Red Fort Complex 8. Hill Forts of Rajasthan 8.1 Jaisalmer Fort 8.2 Amber Fort 8.3 Ranthambore Fort 8.4 Kumbhalgarh Fort 8.5 Chittorgarh Fort 8.6 Gagron Fort 9. The Jantar Mantar, Jaipur 10. Jaipur City, Rajasthan 11. Keoladeo National Park 12. Fatehpur Sikri 13. Agra Fort 14. Taj Mahal 15. Khangchendzonga National Park 16. Manas Wildlife Sanctuary 17. Kaziranga National Park 18. Khajuraho Group of Monuments 19. Archaeological Site of Nalanda Mahavihara at Nalanda, Bihar 20. Mahabodhi Temple Complex at Bodh Gaya 21. Rani-ki-Vav (the Queen's Stepwell) at Patan, Gujarat 22. Buddhist Monuments at Sanchi 23. Historic City of Ahmadabad 24. Champaner-Pavagadh Archaeological Park 25. Rock Shelters of Bhimbetka 26. Sundarbans National Park 27. Ajanta Caves 28. Ellora Caves 29. Elephanta Caves 30. Chhatrapati Shivaji Terminus (formerly Victoria Terminus) 31. Victorian Gothic and Art Deco Ensembles of Mumbai 32. Sun Temple, Konârak 33. Western Ghats 33.1 Chandoli National Park 33.2 Agumbe Reserved Forest 33.3 Pushpagiri Wildlife Sanctuary 33.4 Silent Valley National Park 33.5 Chinnar Wildlife Sanctuary 33.6 Eravikulam National Park (and proposed extension) 33.7 Achankovil Forest Division 33.8 Shendurney Wildlife Sanctuary 34. Group of Monuments at Pattadakal 35. Churches and Convents of Goa 36. Group of Monuments at Hampi 37. Group of Monuments at Mahabalipuram 38. Great Living Chola Temples

Type of Heritage Site: ● Cultural ● Natural ○ Mixed

Figure 1.2 UNESCO World Heritage Sites in India in 2019

(Source: https://whc.unesco.org/en/statesparties/in, accessed September 30 2019)

around the 5th and 6th century CE (www.britannica.com/place/Ajanta-Caves). Elephanta caves are on top of a hill situated on the Elephanta Island in the Arabian Sea. The caves portray Lord Shiva's magnificent sculptures as a creator, destroyer and preserver (www.britannica.com/place/Elephanta-Island).

Hampi in Bellary District of Karnataka, was a prosperous capital city of the Vijayanagara Empire by the banks of Tungabhadra River from the 14th to 17th century. Hampi has 1600 monuments, but they are in ruins now. The monuments include Hindu temples with Hindu deities, Jain temples, mosques, tombs and water tanks. Stone used for these monuments was procured from the abundant granite source around Hampi (www.britannica.com/place/Vijayanagar).

Konark Temple, located in Puri District of Odisha, was also known as the Black Pagoda and was constructed as Giant Chariot, dedicated to the Sun God. The chariot stood on 12 wheels carved in stones with seven horses on the sides and front. It is a World Heritage Site and was built in the 13th century. Exquisite sculptures were carved on the walls of the temple depicting different subjects. The stones used for constructing the different sections of the temple are khondalite, laterite, sandstone, soapstone, serpentine, marble and granite (www.britannica.com/place/Odisha/History).

Chittorgarh Fort is one of the largest forts in Rajasthan and a designated UNESCO World Heritage Site too. It includes temples, palaces, water bodies and cenotaphs of martyred nobles, most of them built in Vindhyan Sandstone (Kaur *et al.*, 2019b; www.britannica.com/place/Chittaurgarh; https://www.chittorgarh.com/article/chittorgarh-history/231/).

1.4 International Union of Geological Sciences (IUGS)

IUGS promotes earth sciences by supporting broad-based scientific activities, advancing geological education and propagating public awareness of geology. The IUGS has various commissions and subcommissions based on the topical themes of geology which are of interest to the geologists in particular and to humankind in general (www.iugs.org/). The International Commission on Geoheritage (ICG) is one such commission of IUGS which deals with geoheritage and heritage stones (http://adsabs.harvard.edu/abs/2017EGUGA.19.2639P). The ICG has two subcommissions viz., Heritage Sites and Collections Subcommission (HSCS) and Heritage Stones Subcommission (HSS). The Heritage Stones Subcommission is primarily engaged in identifying the natural stones which have been instrumental in the stone built architectural heritage around the globe (http://globalheritagestone.com/).

1.4.1 Heritage Stone Subcommission (HSS): Global Heritage Stone Resource (GHSR)

The natural stones have contributed to the magnificent architectural heritage built across the globe, reflecting their unique cultural lineage. The HSS has been encouraging nations to identify their respective heritage stones used in the architectural heritage with cultural and global relevance to be nominated and approved by IUGS as GHSRs. Given in the following is the definition of GHSR for the ready reference of readers from the link (http://globalheritagestone.com/reports-and-documents/terms-of-reference/).

> A **Global Heritage Stone Resource** (GHSR) is a designated natural stone that has achieved widespread use over a significant historical period with due recognition in human culture.

The GHSR designation of a stone can have important implications in the field of geology, engineering, architecture, cultural studies, historical quarries, present day quarries, conservation and restoration of architectural heritage etc. (Pereira *et al.*, 2015b). It also serves the purpose of a great outreach activity outside the geological realm where people are made aware of the existing stone heritage and their involvement on a wider platform to safeguard and take pride in it.

The recognition of a stone as a GHSR involves various steps which are governed by the Terms of Reference (ToR) prepared by the Heritage Stone Subcommission board members and finally approved by the Executive Committee (EC) of the IUGS. The Terms of Reference (2017) are available on the link: (http://globalheritagestone.com/reports-and-documents/terms-of-reference/). The revision of ToR of the HSS is underway and can be found on the website of the Heritage Stone Subcommission. A stone may be designated a GHSR if the following criteria are met by the stone given in the link for the ready reference of readers (http://globalheritagestone.com/reportsand-documents/terms-of-reference/):

1. Historic use for a period of at least 50 years
2. Wide-ranging geographic application
3. Utilisation in significant public or industrial projects
4. Common recognition as a cultural icon, potentially including association with national identity or a significant individual contribution to architecture
5. Ongoing availability of material for quarrying
6. Potential benefits (cultural, scientific, architectural, environmental) arising from GHSR designation

To get the stone the status of Global Heritage Stone Resource one needs to prepare a nomination in the form of an article with acceptance in a high impact factor research journal. The publication in a reputed international journal is promoted to cover readers across the globe and give the stones their due recognition. The published article on the proposed stone along with other supporting publications and references are sent for evaluation to the HSS board where they are scrutinized by the board members of the HSS with the help of its advisory members and correspondents taking into account the ToR of the subcommission. If, the consensus among the board members prevail on the satisfactory report of the evaluators, a complete summary of the proposed stone is prepared by the Secretary General (SG) of HSS and sent to the IUGS EC. The IUGS executive on the basis of the report prepared by the HSS SG and other supporting documents takes the final decision of ratification of the nominated stone as a GHSR.

1.4.2 GHSR: global scenario

To date, 22 heritage stones have earned recognition by the IUGS as Global Heritage Stone Resource (Fig. 1.3; Table 1.2; http://globalheritagestone.com/other-projects/ghsr/designations/). Each heritage stone has widespread historical use along with features like appealing texture, sustainability, versatility in use reflecting cultural ethos unique to each place, international use, functional historical quarries or conserved nonfunctional historical quarries. Most of the designated GHSRs were not only used for ancient architecture but are still in huge demand and find utility in the contemporary world too. The majority of the stones recognized as GHSRs till September 2019 come from Europe, two from North America and one from South America and Asia, each (Fig. 1.3; Table 1.2; http://globalheritagestone.com/otherprojects/ghsr/designations/). These stones were used to create valuable architectural heritage in the past and are still revered and cherished by the current generation. The grandeur of the stones can be reflected in robust work done by sculptors, architects and artisans. IUGS is making a tremendous effort at recognizing the stones from around the world for their heritage value. However, there are still innumerable deserving stones that were used in the architectural heritage which surely need to be added to the current list of GHSRs. In the following is a brief account of stones already designated as GHSRs by IUGS-HSS.

Carrara marble from Italy has a quarrying history of almost 2000 years which started during Roman times. Carrara marble has been a favorite marble of artists and architects such as Michelangelo, Jacopo Della Quercia and Bernini, to name a few. Many public buildings, architectural

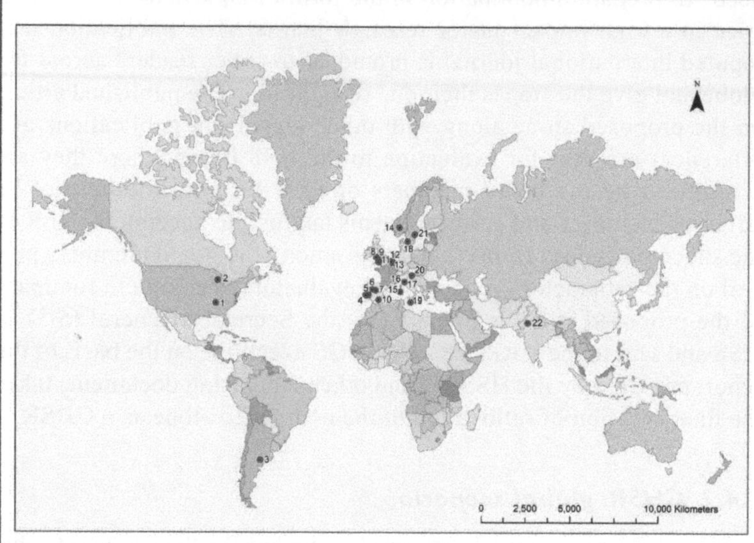

IUGS-HSS designated GHSR Sites on world map

1. Tennessee Marble, USA 2. Jacobsville Sandstone, USA 3. Piedra Mar Del Plata, Argentina 4. Lioz Stone, Portugal 5. Estremoz Marble, Portugal 6. Villamayor Stone, Spain 7. Alpedrete Stone, Spain 8. Welsh Slate, United Kingdom 9. Bath Stone, United Kingdom 10. Almeria, Spain 11. Portland Stone, United Kingdom 12. Lede Stone, Belgium 13. Pierre Bleue Rock, Belgium 14. Larvikite, Norway 15. Rosa Beta Granite, Italy 16. Pietra Serena, Italy 17. Carrara Marble, Italy 18. Hallandia Gneiss, Sweden 19. Lower Globigerina Limestone, Malta 20. Podpec Limestone, Slovenia 21. Kolmarden Stone, Sweden 22. Makrana Marble, India

Figure 1.3 IUGS-HSS designated GHSR sites on the world map in 2019 (Source: http://globalheritagestone.com/igcp-637/igcp-achievements/, accessed September 30 2019)

structures and sculptures have used Carrara stone for its utilitarian value, and it is exported all over the world (Primavori, 2015).

Lower Globigerina limestone is soft yellow stone from Malta. This stone has been put to use from prehistoric times, and most of the buildings and sculptures in Malta have used this stone as a major building material. It has calcite as a major component. It has a historical use dating back 6000 years with great cultural value. It is still quarried, has international use and is exported to different places in Europe, North Africa, Greece and the UK (Cassar *et al.*, 2017).

Table 1.2 List of 22 GHSRs designated from all around the globe in 2019

Sr. No	*GHSR*	*Place*	*Rock type*	*Historical and current use (national and international)**	*Reference/ citation*
1	Jacobsville sandstone	Michigan, USA	Sandstone	Cotton Exchange, New Orleans Masonic Temple, Chicago John Munro and Mary Beecher Long year House, Michigan Red Jacket Town Hall and Opera House, Michigan	Rose et *al.*, 2017
2	Tennessee marble	Tennessee, USA	Marble	U.S. Custom House, Post Office, Federal Building, Knoxville, Tenn. The J.P. Morgan Library, New York, USA Tennessee Supreme Court Building, Nashville, Tenn. National Gallery of Art, Washington, D.C., USA U.S. Capitol, Washington, D.C., USA Manitoba Legislative Building, Winnipeg, Canada Union Station, Toronto, Canada	Byerly and Knowles, 2017
3	Piedra Mar Del Plata	Mar del Plata, Argentina	Quartzite	Santa Cecilia Chapel, Argentina Stella Maris Church, Argentina General Pueyrredon City Hall, Argentina Monk's Tower, Argentina Opera Theatre, Argentina Sea lions, Argentina Diagonal, Argentina Norte and Florida, Argentina	Cravero et *al.*, 2015
4	Estremoz marble	Estremoz, Alentejo Province, Portugal	Marble	Roman Temple in Évora, Portugal The Roman Theatre in Mérida, Spain The Roman towns of Ammaia, Portugal Volubillis, Morocco	Lopes and Martins, 2015

(*Continued*)

Table 1.2 (Continued)

Sr. No	GHSR	Place	Rock type	Historical and current use (national and international)*	Reference/ citation
5	Lioz stone	Lisbon, Portugal	Limestone	Jesuit Church of São Roque, Portugal São Jorge Castle in Lisbon, Portugal The Convent of Christ, Portugal Baroque Mafra Monument, Portugal Salvador in Bahia, Brazil	Silva, 2019
6	Villamayor stone	Salamanca, Spain	Sandstone	Old Cathedral and San Julian church, Salamanca Gothic monuments (Spanish plateresque style) such as the New Cathedral and the Church of San Esteban, Salamanca Sculpted facade of the Salamanca University of Salamanca (one of the oldest Universities in Europe) Galleries and arcades of the sumptuous Baroque Main Square of Salamanca, Private mansions of Salamanca	Garcia-Talegon *et al.*, 2015
7	Alpedrete granite	Alpedrete Province, Madrid, Spain	Granite	Palace of Marid, Spain Alcalá Gate, Spain The National Library, Spain Artichoke Fountain, Prado Museum	Freire-Lista *et al.*, 2015
8	Macael marble	Almeria, Spain	Marble	Phoenician sarcophagi, Spain The paleo-Christian Tomb of Berja Almería, Spain Cathedral-Mosque of Cordoba, Spain Cathedral and the Royal Chapel in Granada, Spain Royal Palace of Madrid, Spain	Navarro *et al.*, 2019

9	Welsh slate	Wales, United Kingdom	Slate	Buckingham Palace, London, UK Cathedral House, Glasgow, UK Jesus College, Cambridge, UK Kunsthistorisches Museum Vienna, Austria St. James' Anglican Church, Sydney Australia Amalienborg (The Royal House), Copenhagen, Denmark Hotel de Ville, Paris, France Royal Mausoleum, Hawaii Trinity College Dublin, Ireland Arts Centre, Christchurch, New Zealand The English Church, Gothenburg, Sweden The Red House, Trinidad, West Indies	Hughes *et al.*, 2016.
10	Port and stone	Portland, United Kingdom	Limestone	St Paul's Cathedral, United Kingdom Palace of Westminster, UK First stone London Bridge, UK British Museum, UK Bank of England, UK United Nations building in New York City, USA National Gallery of Ireland, Dublin, Ireland Parliament Building, Dublin, Ireland Casino Kursaal, Ostend, Belgium Villa at Neshua, Kuwait Chubu Electric Building, Japan Zagaleta project, Andalusia, Spain	Hughes *et al.*, 2013
11	Bath stone	Bath, United Kingdom	Limestone	Queen Square, Bath, UK Royal Crescent Bath, UK Pultney Bridge, Bath, UK	Marker, 2015

(*Continued*)

Table 1.2 (Continued)

Sr. No	*GHSR*	*Place*	*Rock type*	*Historical and current use (national and international)**	*Reference/ citation*
				York street, Bath, UK Apsley House, London, UK Royal Pavilion, Brighton, UK Temple Meads Railway Station, Bristol, UK	
12	Petit granit/ Pierre Bleue rock	Namur, Belgium	Limestone	Cathedral of Funchal, Madeira Collegiate church of Sainte-Waudru, Mons, Belgium Garden pavilion, parc d'Arenberg, Enghien, Belgium Castle of Seneffe, Belgium Brussels airport, Belgium Leipzig airport, Germany Chartres town center, France Design center, Saint-Etienne, France The Hague town hall, Netherlands	Pereira *et al.*, 2015a
13	Lede stone	Brussels, Belgium	Sandy limestone	St Michael's Church, Belgium St Walburga Church, Belgium House of the Free Boatsmen, Belgium Guildhouse of the Bricklayers, Belgium St Peter Railway station in Ghent City Hall, Belgium St-Lievens Monster Tower, Netherlands Church of Our Lady, Netherlands	De Kock *et al.*, 2015

14	Larvikite	Larvik, Norway	Monzonite	Art Nouveau buildings, Norway Harrods in London, UK Galleries Lafayette in Paris, France Art Deco buildings in London, UK	Heldal *et al.*, 2015
15	Hallandia gneiss	Getinge, Sweden	Gneiss	Mostorp Manor, Sweden Hjuleberg Manor, Sweden Parliament house, Sweden Streets, squares, Denmark and Poland Peace monument Reutersward, The Hague President Washington monument, USA	Schouenborg *et al.*, 2015
16	Carrara marble	Tuscany, Italy	Marble	Temple of Apollo Palatinus, Italy Michelangelo's Pieta, Italy Miracle Square, Baptistere, Cathedral and Leaning Tower, Italy Pantheon, Italy Oslo Opera House, Norway Sheikh Zayed Mosque, Abu Dhabi, UAE Akshardham, India	Primavori, 2015
17	Pietra Serena	Florence, Italy	Sandstone	Santa Croce Convent, Florence, Italy Franciscan Convent of Santa Croce, Florence, Italy The Spedale degli Innocenti, Florence, Italy Zuccari Palace-Florence, Italy	Fratini *et al.*, 2015
18	Rosa Beta granite	Sardinia, Italy	Granite	Pantheon in Rome, Italy Renaissance Cathedral in Pisa, Italy Financial Square Building, New York City, USA	Careddu and Grillo, 2015
19	Lower Globigerina Limestone	Malta	Limestone	UNESCO-listed Prehistoric Megalithic Temples of the Maltese Islands Fortified cities of Valletta, Mdina, Malta Citadel of Gozo UNESCO building, Paris, France	Cassar *et al.*, 2017

(*Continued*)

Table 1.2 (Continued)

Sr. No	*GHSR*	*Place*	*Rock type*	*Historical and current use (national and international)**	*Reference/ citation*
20	Podpec limestone	Podpec, Slovenia	Limestone	Central Stadium in Ljubljana, Slovenia The Faculty of Natural Sciences and Technology of the University of Ljubljana, Slovenia Ljubljana's City Hall, Slovenia Parts of the Parliament of the Republic of Slovenia Antonius Church, Serbia	Kramar *et al.*, 2015
21	Kolmarden stone	Kolmarden, Sweden	Serpentine marble	Drottningholm Castle, Sweden Natural History Museum, Sweden Matchstick Palace, Sweden Stockholm Town Hall (where the Nobel prize reception takes place), Sweden Leeds University Library, UK Copenhagen Town hall, Denmark Paris Opera house, France	Wikström and Pereira, 2015
22	Makrana marble	Makrana, India	Marble	Taj Mahal, India Victoria Memorial, India Birla Mandir, Jaipur, India Pearl Mosque, Pakistan	Garg *et al.*, 2019

* A few iconic monuments are listed. For details refer to the original papers cited in the last column.

(Source: http://globalheritagestone.com/other-projects/ghsr/designations/, accessed September 20 2019)

Hallandia gneiss has documented records of quarrying since 1850 when it was exported to Denmark and Germany for paving purposes although it was in local use much prior to that. It is a gneiss which is aesthetically appealing owing to textures imparted by veining. There were 500 operational quarries reported in the past, but the quarrying in present times is minimal. It is imperative to preserve the quarries for future restoration of the monuments (Schouenborg *et al.*, 2015).

Jacobsville sandstone is popular red sandstone from Michigan, USA. Numerous buildings in Michigan and eastern North America constructed during late 19th and early 20th centuries used Jacobsville sandstone. The well-known Waldorf Astoria Hotel in New York was made using this red sandstone. Quarrying in the vicinity of the Jacobsville, Michigan plateau was reported from 1880 to 1920. At present there are no functional quarries near Jacobsville (Rose *et al.*, 2017).

Kolmarden stone is a green serpentine marble from Sweden. The use of Kolmarden stone goes back to the 13th century. It has been designated as a GHSR due to its widespread usage in historical buildings of Sweden and elsewhere in Europe. To date, quarrying of Kolmarden marble continues in a small area (Wikström and Pereira, 2015).

Larvikite is an unusual monzonite with blue iridescent feldspars imparting an attractive appeal to the rock. The rock has been used in Norway since the 12th century. It became very popular on a global platform from 19th century onwards for ornamental and monumental work. Currently, it is being quarried on a large scale in a sustainable manner and is exported around the globe (Heldal *et al.*, 2015).

Lede stone is a sandy limestone from Belgium. This stone is associated with the Belgian and Dutch cultural heritage. The use of this stone has been dated back to the Roman period. Lede stone was an important stone in the Gothic architecture in the Middle Ages. As of now, only one functional quarry is producing Lede stone which is primarily used for restoration purposes (De Kock *et al.*, 2015).

Lioz stone is microcrystalline limestone from Portugal. It is a fossiliferous limestone which imparts aesthetic appeal to the rock and has been in use since the 16th century. It has been recognized as the "Royal stone" for its use in a lot of stately buildings. It is still being quarried and also used in the restoration of old buildings (Silva, 2019).

Piedra Mar Del Plata is orthoquartzite from Argentina. It has been in use for more than 50 years as a building material. It is a durable and affordable stone thus making it popular amongst the locals for construction of their abodes in Argentina. Initially, it was used in the city of Mar del Plata (beach resort) for construction in a distinct architectural style, but now it is used in other parts of the country making its architectural use heritage of Argentina (Cravero *et al.*, 2015).

Podpec limestone is a dark grey to black color carbonate rock from Slovenia. The limestone has fossil shells thus making it an attractive rock. Its usage can be dated from Roman times to the present. The main quarry near the village of Podpec, which provided stones for most of the historical building, is no longer functional, and it has been assigned the status of natural monument and is a protected site (Kramar *et al.*, 2015).

Portland stone is ooidal limestone from the United Kingdom. This stone is also said to have found use since the Roman times and was a preferred choice of architects for building cathedrals and other monuments in the medieval period. This stone is currently quarried, and many modern new buildings have been constructed using Portland stone. It is exported outside the UK, and many international buildings are made using Portland stone (Hughes *et al.*, 2013).

Villamayor stone is golden sandstone from Spain. Salamanca, a UNESCO World Heritage City of Spain, has many Romanesque, Gothic and Baroque style buildings with facades done primarily in Villamayor stone. Some families are still in the quarrying business of this stone which is mostly used for restoration purposes of old buildings. It is important to safeguard the historical quarries of Villamayor stone for future generations (Garcia-Talegon, 2015).

Welsh slate is a well-known source of building material from the United Kingdom. Its use has been documented since the Roman period in Wales. It is primarily used for roofing besides cladding and monumental masonry. It is used as a restoration material for many heritage buildings. It is currently quarried for building purposes and is also exported to many European countries (Hughes *et al.*, 2016).

Petit granite or Pierre Bleue rock is grey-bluish crinoidal limestone from Belgium. It has been in use since the Middle Ages. The use of this rock became more prevalent from the 16th century onwards. The quarrying of Pierre Bleue is ongoing and adding to the stone economy of Belgium. The stone is popular in contemporary architecture of Brussels and other cities of Belgium. It is exported worldwide for its building material and aesthetic value (Pereira *et al.*, 2015a).

Pietra Serena is sandstone from Florence, Italy. It was the most important material used in Florentine Renaissance architecture. It is a sandstone with calcitic cement. This stone has been quarried from hills in the vicinity of Florence for centuries and was a popular building material. Its use lasted till the 19th century, and currently no quarrying is taking place due to depleting resource and the high cost of the stone (Fratini *et al.*, 2015).

Macael marble is from Spain. It has been in use for more than 5000 years. The most famous variety is white Macael (Blanco Macael). Macael marble is in great use nationally and internationally. It has active quarries, and the marble is exported all over the world (Navarro *et al.*, 2019).

Bath stone is limestone from England. The records of its use can be dated back 2000 years in terms of local use but the widespread use of this limestone is reported for the past 350 years. This limestone has been used extensively in the United Kingdom and elsewhere to some extent. Bath city, which has architectural heritage built in this limestone, is a UNESCO World Heritage City. Bath stone is presently quarried to meet the demands of the dimension stone industry and has a long way to go (Marker, 2015).

Tennessee marble is limestone from Tennessee, North America. Tennessee stone has been quarried for the dimension stone industry and lime since the Colonial times in America. The quarrying of Tennessee can be worked out in relation to the construction of the Francis Alexander Ramsey home in Knox County, Tennessee, USA, in 1797. The stone is a popular dimension stone in USA and Canada and is being quarried for sculptural purposes as well (Byerly and Knowles, 2017).

Estremoz marble is marble from Portugal. This marble has been in use for architectonic creations since antiquity. The use of Estremoz marble goes back to the 4th century BC, and it finds use in the current scenario as well. It is exported to almost all European nations, China, Brazil, Egypt and Taiwan (Lopes and Martins, 2015).

Alpedrete granite is a monzogranite from Spain. Its usage dates back to the Neolithic period. It was mostly quarried from Alpedrete and its surroundings in the past. The current active quarry sites have numerous historical quarry sites in the vicinity. Alpedrete granite adorns almost 75% of the culturally significant architectural heritage of Madrid (Freire-Lista *et al.*, 2015).

Makrana marble is calcitic marble from Makrana, India. Makrana marble is a unique marble which gave character and essence to the world famous Taj Mahal, also listed as one of the modern Seven Wonders of the World. The marble from Makrana has been quarried for more than 400 years. The marble is still quarried from Makrana although the quarrying has become unsafe due to the increased depth of the marble resource. The Makrana marble is used for a variety of purposes such as cladding, flooring and sculpturing. The historical quarries of Makrana need to be preserved for future restoration work of the heritage monuments (Garg *et al.*, 2019).

1.4.3 GHSR: Indian scenario

India records architectural heritage built in stones from time immemorial such as the Bhimbetka caves of Madhya Pradesh in Vindhyan Sandstone, the Ajanta and Ellora in-situ cave temples in Maharashtra engraved well in Deccan basalt, Khajuraho temples of Madhya Pradesh carved in Vindhyan Sandstone and Konark group of temples ornately carved in Khondalite, to name a few (Guha and Roonwal, 2014). India deserves to have a lot more stones designated as GHSRs owing to its diverse and umpteen architectural heritages framed in stone. As of now, India has one stone i.e. Makrana Marble from Rajasthan designated as a GHSR (Garg *et al.*, 2019; www.hindustantimes.com/punjab/marble-used-for-taj-mahal-is-now-global-heritage-stone-resource/storyt67WWKE5kj05JL9o-3qEdlO.html).

In contrast to the European GHSRs, India is way behind and needs to step up promoting its heritage stones in the global arena. To date, three important stones from India have been proposed following the criteria of designation of GHSR, namely, Deccan basalt and trachyte (Kaur *et al.*, 2019a) and the Vindhyan Sandstone (Kaur *et al.*, 2019b). Hopefully in the near future, Indian geoscientists will propose more stones at international platforms to be designated as GHSRs owing to India's rich stone built architectural heritage.

Chapter 2

Delhi and Agra vis-à-vis monuments

> *"The history of India is traced as clearly as possible upon its monuments. These last unfortunately are disappearing with regrettable rapidity."*
>
> Gustave le Bon

2.1 Introduction

India was a land of abundant resources, rich cultural and architectural heritage in ancient times, hence accorded the moniker "the Golden Sparrow." India was a leader in trade, economic understanding and spirituality coupled with a bountiful share of rich metals, precious stones and spices. It was far more expansive as compared to its current outline on the political map and was strategically located, with its fair share of natural boundaries all around that helped in keeping it safe from any impending foreign attacks. These natural boundaries were a challenge for any invader and necessitated extensive military acumen to enter India. India's wealth and abundant resources enticed rulers to embark from different destinations across the globe for centuries. They either entered India via the high mountains up north, water channels or deserts (www.britannica.com/place/India/History-ref484915).

The focus of this book is on the World Heritage Sites of Delhi and Agra and the stones used for their construction. The cities of Delhi and Agra together unfold a significant narrative of medieval Indian history. The monuments, under the ambit of the present study, highlight crucial milestones that encompass the initiation, growth and takeover of the Mughal Empire. The tail end merges with the first signs of Colonial intervention.

2.2 Delhi and its monuments

> " '*I asked my soul: What is Delhi? She replied: The world is the body and Delhi its life.*'
> *Mirza Asadullah Khan Ghalib*"
>
> Khushwant Singh (*Delhi: A Novel*)

The several invaders and crusaders, who came to Delhi from distant lands to make their presence felt, stayed here, because India was a land of opportunities and fulfilled their ambition for material wealth, propagating religious values amongst the natives and expansion of their kingdoms. They left their legacy in the form of architectural heritage, culture, art, literature, language and religion. This was also the main reason for Delhi to turn into a land of diversity which accepted everyone in addition to an inner resilience.

There are 1208 monuments listed in Delhi (www.intachdelhichapter.org/listings.php) which belong to different time periods, and the majority of them were constructed using sandstone, marble and quartzite. Archaeological Survey of India (ASI) recognizes 174 monuments, which includes three UNESCO World Heritage Sites. The monuments have withstood the vagaries of time, nature (weathering due to rain, heat, wind and moss), urbanization, pollution and vandalization, too. The monuments under ASI are renovated regularly for any damage such as cracks, broken loose blocks and discoloration of stones, in addition to guarding the heritage monuments for longevity and sustenance. Some of the monuments that are not protected by ASI are in a dilapidated state and some have totally disappeared either due to vandalizing, climatic changes, pollution or expansion of the city.

Delhi records several prior cities built and rebuilt by different dynasts of Delhi who came up with a whole new architecture to address their political, religious, cultural and social needs. A peep into the history of Delhi from the 11th century onwards reveals the presence of nine cities, which existed in the Delhi triangle formed between the Delhi Ridge and the river Yamuna (Fig. 2.1, 2.2 and 2.3). Some historians and archaeologists claim that there were seven cities that existed (Spear, 1943), but in this book, a brief account of nine cities has been discussed starting from the 11th century onwards that came together to make the present day Delhi which we call the contemporary 10th city (Fig. 2.3). Lal Kot and Qila Rai Pithora are discussed as two different cities in this book contrary to most narratives which describe them together as the first city of Delhi (Hearn, 1906; Spear, 1943; Gupta, 1981; Balasubramaniam, 2005; Steven, 2019).

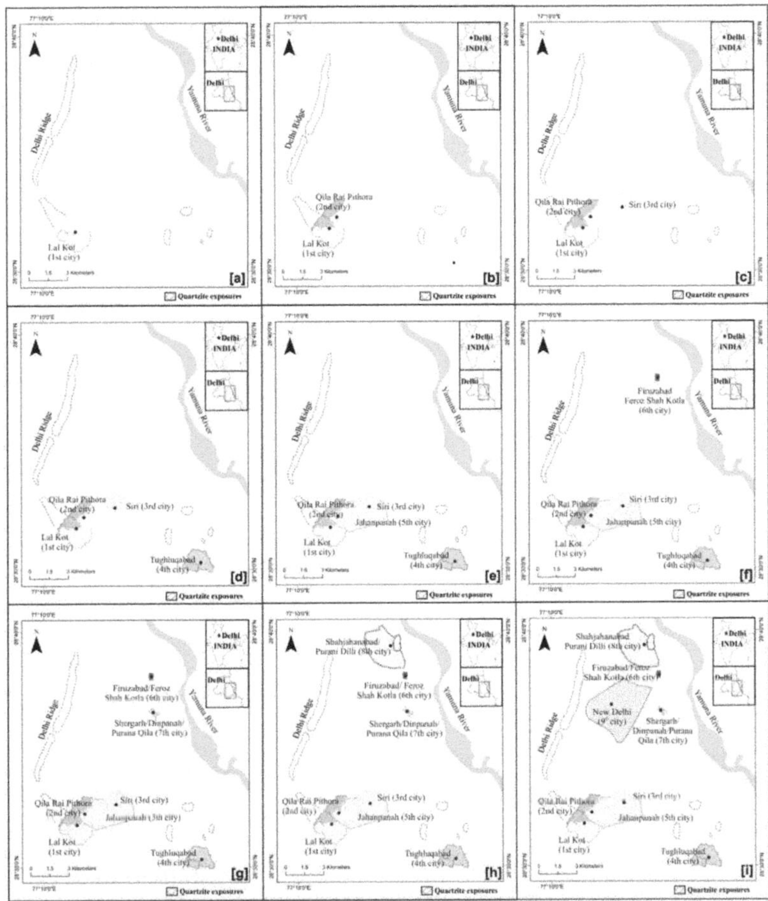

Figure 2.1 Course of development and spread of cities in Delhi:(a) Lal Kot; (b) Qila Rai Pithora; (c) Siri; (d) Tughlaqabad; (e) Jahanpanah; (f) Firuzabad; (g) Shergarh/Dinpanah; (h) Shahjahanabad; (i) New Delhi

2.2.1 First city: Lal Kot

The first city of the 11th century, Lal Kot, was constructed in red sandstone under the leadership of Tomar King, Anangpal (Fig. 2.1a). The strategic location of the fort in the rocky terrain of Aravalli hills gave it the upper

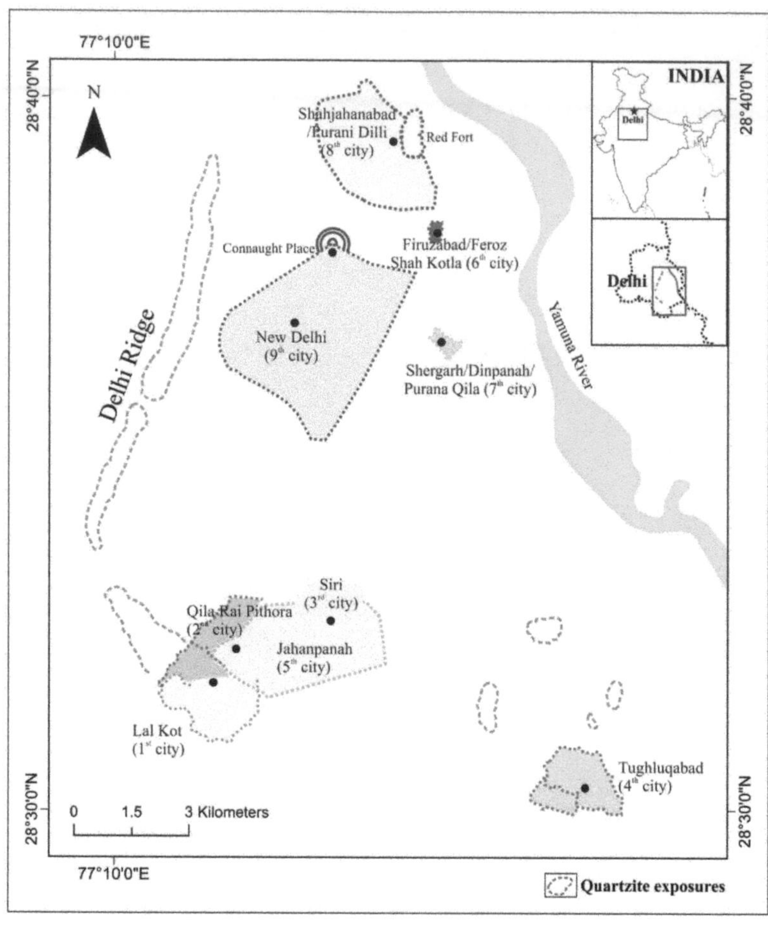

Figure 2.2 Map of Delhi showing extensions of nine cities established since 11th century

hand in terms of guarding the city and keeping it safe. Tomars also built the Anangpur dam and *baolis*. The dam is 2 km away from the Surajkund water reservoir in Faridabad and made in quartzite. It is 50 m long and 7 m high, with a fine structure in place to control the flow of water (Sharma, 2001; Singh, 1996; http://nmma.nic.in/nmma/nmma_doc/IndianArchaeology Review/Indian Archaeology 1991-92 A Review.pdf).

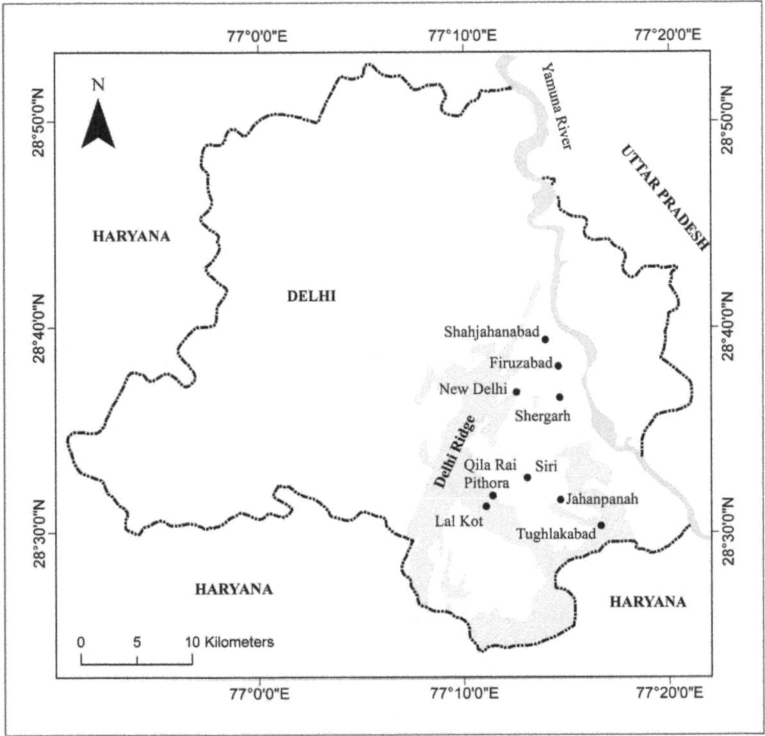

Figure 2.3 Map showing compiled locations of nine cities within present day Delhi

2.2.2 Second city: Qila Rai Pithora

After the Tomars, Prithviraj Chauhan ruled from Lal Kot and the fort was extended in span to be known as Qila Rai Pithora in the 12th century. This city flourished under the rule of Prithviraj Chauhan (Fig. 2.1b). "Qila" means fort and it is known for its massive ramparts, a requirement at the time to safeguard the city (www.britannica.com/place/Delhi/History-ref293596). The Qutb Minar and its adjoining monuments, commonly referred to as Qutb Complex, are within the precincts of the second city. The Tomars and Chauhans built a number of Hindu temples during their rule that were supposedly destroyed by Mohammad Ghori and eventually became part of the Quwwat-ul-Islam mosque in the Qutb

Complex (Page and Sharma, 2002). The ruins of Qila Rai Pithora can still be seen in the present day Saket, Mehrauli, Kishangarh and Vasant Kunj areas of Delhi. Qila Rai Pithora is in ruins, and all one gets to see are the mounds of the past fortification.

The Delhi Sultanate began with the advent of Mahmud of Ghazni followed by Mohammed Ghori who battled and defeated the Rajput king, Prithviraj Chauhan. These invaders vandalized the ancient Hindu temples and monuments. Then came Qutbuddin Aibak from the slave dynasty and initiated the construction of Qutb Minar (a towering monument) which is now part of Qutb Complex, a UNESCO Heritage site of Delhi. Due to his untimely death, his son-in-law, Iltutmish, completed the tiers of the minaret (Page and Sharma, 2002). The Qutb Complex went through repair and renovation several times during the reign of different Sultanate rulers who extended the complex by adding more features like a Quwwat-ul-Islam mosque, Alai Darwaza, Iron Pillar, sandstone screen, tombs and Alai Minar. The Qutb Minar and adjoining monuments were mainly constructed in red sandstone, quartzite, marble and slate.

2.2.3 Third city: Siri

Alauddin Khalji, the second ruler of the Khalji dynasty, built the third city of Delhi called Siri between 1297 and 1307 (Fig. 2.1c). It is also the first city constructed from scratch by the Islamic Sultanate ruler. He was successful at keeping the rampant Mongol attacks at bay. He made Siri his capital, and it was in close proximity to Qutb Minar in the northeast direction. The Islamic architecture was in its formative years, and Alauddin used the skills of refugee Seljuk artists to contribute to the architecture of the city. In 1398 AD, Timurlane, a Mongol ruler wrote in his memoirs, "The Siri is a round city with enormous buildings surrounded by strong fortifications made of stone and brick." The city was oval in shape, with palaces and other structures. There were seven gates to the city for entry and exit. The palace had a thousand pillars (made of wood) and hence was known as "Hazar Sutan" (Spear, 1943; Peck, 2005). The palace was erected outside of the fort and the flooring was entirely done with marble, and other stones were used for ornamental value. The doors were carved and embellished with precious stones. The fort is in a dilapidated state now, and one gets to see only the ruins of fort walls in Shahpur Jat village. Alauddin also got the water reservoir constructed at Hauz Khas to fulfill the water supply at the Siri Fort. He made considerable changes to the Quwwat-ul-Islam mosque and made it four times bigger than what his predecessor had left. He also got the construction of Alai Minar started with the intent of outdoing the size of Qutb Minar. His dream was not fulfilled as he died soon after the construction of the first storey. It still stands in rubble and reminds one of his unfulfilled wish.

2.2.4 *Fourth city: Tughlaqabad*

After the Khaljis, the reigns of Delhi Sultanate were passed on to the Tughlaq dynasty. Ghiyasuddin Tughlaq built the fourth city Tughlaqabad in 1321 (Fig. 2.1d). He was a learned man, comfortable with Arabic and Persian languages. Ghiyasuddin Tughlaq initiated Tughlaqabad fort primarily for resisting the Mongol attacks. It was made on the Aravalli's rocky hills. The fort was made in quartzite and sealed with mud mortar. The stone was probably procured from hills in close proximity. The craftsmanship exhibits the Islamic architecture used for different areas of the fort, like the parapets, bastions, huge gates, palaces, mosques and tombs with thick walls and high fortification. It is believed that the city had 52 gates, but only a few of them exist at present.

2.2.5 *Fifth city: Jahanpanah*

Mohammad Bin Tughlaq built the fifth city Jahanpanah (literally, "Refuge of the world"), which constituted a palace for the royal family and the subjects inside the walls of the fort (Fig. 2.1e). Jahanpanah fort was built in a very large area and presently is in ruins. It had several monuments in the precinct such as tombs, mosques, palaces and other structures made in red sandstone and marble (Peck, 2005). Bin Tughlaq tried to absorb the earlier cities like Qila Rai Pithora and Siri with Adilabad fort as a measure to control the rampant Mongol invasions. The Moroccan scholar and traveler Ibn Batuta's records mention Bijay Mandal which was a palace with 1000 pillars, also known as "Hazar Sutan." It is an octagonal two-storey structure, which gives a view of the city around. This monument is also linked to Alauddin Khalji. Adilabad fort was constructed by Mohammad Bin Tughlaq in the 14th century, after the death of his father, Ghiyasuddin Tughlaq. It faces the Tughlaqabad fort and was very similar in architecture and use of building materials to the former one. Even though the fort is one of the oldest and has been in the city for all these years, it was not frequented much. The inside has some palace walkways; underground cells used as grain bins. The precincts of Jahanpanah are now part of the urban expansion around South Delhi in localities like Adchini, Panchsheel, Malviya Nagar and Aurobindo Asharam.

2.2.6 *Sixth city: Firuzabad*

Firoz Shah Tughlaq founded the sixth city Firuzabad in 1351 by the banks of the river Yamuna (Fig. 2.1f). This city stretched from the Northern Ridge until the Hauz Khas in South Delhi. The Firoz Shah Kotla fort of Firuzabad was built of sandstone. It housed palaces, armories for weapon storage, halls for private and public meetings, quarters for the subjects and servants, barracks for the army, a mosque for worship, *baolis* (the water reservoirs), and royal and public bathing areas in the precincts. The main highlight of the fort was the Ashokan pillar from the time of the Mauryan Empire, which was brought from present day Haryana to be placed near the mosque. This fort did not have huge gates. This city was more to the north of Delhi and drifted from the previous five cities. The Khirki, Kalusarai and Sarai Shahji Mahala mosques were built during this time. The forts and other monuments discussed earlier are in ruins, and it is believed that the building materials from these demolished monuments were used to construct cities like Shergarh and Shahjahanabad.

2.2.7 *Seventh city: Shergarh/Dinpanah*

Shergarh, the seventh city, was founded by an Afghan ruler Sher Shah Suri (Fig. 2.1g). He defeated Humayun, the second ruler of Mughal Empire. The Suris ruled from 1540 to 1556 AD. Abul Fazl, the author of *Akbarnama*, states in his writing that Humayun started building a city called Dinpanah, which means "Refuge of the faithful," on the ruins of Indraprastha, by the banks of the river Yamuna. The project could not be completed as Sher Shah defeated him and exiled him for about 15 years. Sher Shah made changes to the fort by strengthening the walls of the fort for resisting the Mongol attacks and named it Purana Qila. Historians believe that Purana Qila was the site for Indraprastha from where the Pandavas of Mahabharata ruled (Smith, 2010; www.thehindu.com/news/cities/Delhi/Delhis-Mahabharata connection/article16837830.ece).

This fort had three huge gates viz., "Talaq Darwaza" or the "Forbidden gate" in the north, "Humayun Gate" in the south and "Bada Darwaza" in the west. The western gate was the entrance to the fort. These were all made in red sandstone (https://ducic.ac.in/cdn/ducic/

NewsEventsCommons/Gates of Delhi.pdf). Sher Mandal in Purana Qila is an octagonal, two-storey structure made in red sandstone. Some historians believe that Sher Shah constructed it as a pleasure spot while others believe it to be Humayun's library (a retained structure from Dinpanah made by Humayun) with narrow quartzite steps (https://nroer.gov.in/55ab34ff81fccb4f1d806025/file/57d9439f16b51c0da00db599). Humayun, after coming back to power in 1555, moved to the old fort (Purana Qila) and ruled from there. He did not survive for long as he slipped from the steps of Sher Mandal and died in 1556. Humayun's Tomb was built in the latter half of the 16th century when the Mughals shifted their capital to Agra and his son Akbar was focusing on building his empire. This exemplary, significant monument was the first garden tomb in the Indian subcontinent. Its construction work was initiated by his wife Haji/Bega Begum and majorly funded by Akbar. It was constructed on the banks of the river Yamuna under the supervision of Persian architect Mirak Ghiyathuddin. The tomb is also known as the "dormitory of the Mughals" as around 150 Mughals from the royal family and ranks were buried in the garden tomb. The use of red sandstone, marble cladding and inlays for borders, intricate marble jalis (perforated screens) and corbelled brackets define the Mughal architecture (Asher, 1992). The emperor's cenotaph is made in white marble and is situated in the octagonal chamber. The flooring is done in black stone and white marble, arranged in a pattern. It has arched openings and carved stone jali work (Shaikh, 2015).

2.2.8 Eighth city: Shahjahanabad

Shahjahan, the fifth emperor of the Mughal dynasty (Gupta, 1981), founded the eighth city called "Shahjahanabad" in 1649 (Fig. 2.1h). Mughal architecture was reaching its peak in those times. The city had 14 huge gates. When he shifted his capital from Agra to Delhi, he had the Red Fort and the Jama Masjid constructed in Shahjahanabad. The Red Fort complex has several spectacular monuments. Lahori Gate, being the main entrance to the fort on the west wall of the fort, led to Chhatta Bazaar, which was a bazaar with a roof. The buildings inside the Red Fort complex included: Diwan-i-Am, a meeting place or a hall of public audience made in red sandstone; and Diwan-i-Khas, a hall of private audiences meant for the nobles and esteemed courtiers from the emperor's court done in white marble. Moti Masjid, a prayer house, was made in white marble for the inhabitants of the fort. The Sawan Bhadon Pavilions were made in white marble having gardens around the structures. A Nahr-i-Bihisht flowed through the royal quarters and the two pavilions. Mumtaz Mahal, Khas Mahal, Rang Mahal and Hira Mahal, the palaces in the precinct of the fort, were made mostly in marble. The fort experienced a setback during the 1857 mutiny and the British ransacked all the precious belongings from the monument.

2.2.9 *Ninth city: New Delhi*

British rulers built the ninth city, New Delhi, and treated it as new capital of India from December 1911 to 15 August 1947 (Fig. 2.1i). Two renowned British architects were assigned the task of designing the Viceroy's palace and Secretariat on Raisina Hill. Edwin Lutyens designed the Viceroy's palace, and Herbert Baker was the architect of the Central Secretariat. The construction of the massive structures broadly occurred between 1912 and 1931. The architectural style of the buildings erected during this period exhibit an amalgam of Mughal and Rajput design elements which incorporated the use of domes, chattris and chajjas. The structures were made in red sandstone. Post-independence, the Viceroy's Palace became the President's House in 1950 when India became a republic.

Imperial Delhi and post-independence Delhi became home to migrants from Pakistan and people from across India. The infrastructure of the city has changed a lot to fend for the needs of its inhabitants. All the prior cities discussed are now woven together with the passage of time but still have a lot to offer about the significant past. Sir Lutyens played an important role in designing and shaping up New Delhi, hence it is also called Lutyen's Delhi.

2.3 Agra and its monuments

The city of Agra is situated on the banks of the Yamuna in the state of Uttar Pradesh. It is part of the Golden Triangle along with Delhi and Jaipur – a cluster of three cities evincing unparalleled tourist interest and influx. Agra is mostly known for its Mughal age monuments, most prominently Taj Mahal, followed closely by the Agra Fort, both having been accorded the "UNESCO World Heritage Site" status. The city traces its origin to the ancient epic Mahabharata when it was called Agravana. The remains of Mankeshvar temple put the origins at 2000 years old. The place is venerated as the site of incarnation of Vishnu as *Parasu-Rama*. It is also connected with the setting of Lord Krishna's many exploits (Latif, 1896). A sustained literary mention of the city comes up in a *qasida* by the poet Masud-bin sad bin Salman with regard to Mahmud bin Ibrahim who is said to have conquered the Agra Fort in the 12th century (Siddiqi, 2008). After the conquest by Mahmud of Ghaznavi in the 11th century, Agra was reduced to a hamlet until Sikandar Lodhi made it his capital in 1506. Under the Lodhi dynasts, the city saw a surfeit of poets, sufis and a high literary culture. His son Ibrahim Lodhi succeeded him in 1517 and ruled till 1526 when he was defeated in the First Battle of Panipat. The city saw a high point in developmental activities under Zahir-ud-din Mohammad Babur who named it his capital and made hitherto unseen architectural structures like his palace in *Chaarbagh*, gardens irrigated by intricate water channels and Persian pavilions. He created the first ever Persian garden called *Aaraam Bagh*/Garden of Relaxation. Agra was hereon taken over by the Afghans, ruling till 1556. After that Agra was to remain the capital of the Mughal Empire until 1648 (Siddiqi, 2008; Latif, 1896).

Agra prospered under the patronage of the Mughal Empire with an extensive plan of expansion and development. It remained the capital of the Mughal Empire through the reigns of Jahangir, Akbar and Shah Jahan, with each emperor leaving a stamp of their individual identity and aesthetic on the city. Akbar consolidated a renewed hold on the town by reinforcing the ramparts of the Agra Fort, adding a new vocabulary to its existing precincts and developing a new mini township in the outskirts of Agra, Fatehpur Sikri. The Agra city was built as a Mughal military encampment in stone. Akbar added unique features like hammams, serais, bridges and mosques of great architectural significance. He promoted book making and translation bureaus and set up a mint in Agra. (Havell, 1904; Hussain, 1937; Siddiqi, 2008).

Foremost, Jahangir built his father's mausoleum in Sikandara. He added many Persian style gardens within the Agra Fort and other locations. Shah Jahan, with his love of marble, gave the city Taj Mahal and several marble monuments within the Agra Fort. The Moti Masjid or "Pearl Mosque," the Jama Masjid or "Great Mosque," and the *Khas Mahal* were completed under this magnificent emperor. He named Agra as Akbarabad after his grandfather. In 1648, he shifted the capital to Shahjahanabad but Aurangzeb reverted to Akbarabad, and allegedly imprisoned his father in a tragic unfolding of a father-son feud. Shah Jahan died in the ramparts of the Agra Fort, a sad figure gazing at the Taj

Mahal, that pinnacle of love and glory that he had both attained in his life and utterly lost (Havell, 1904).

Other noteworthy monuments in Agra City apart from Taj Mahal and Agra Fort that deserve a mention are (Fig. 2.4):

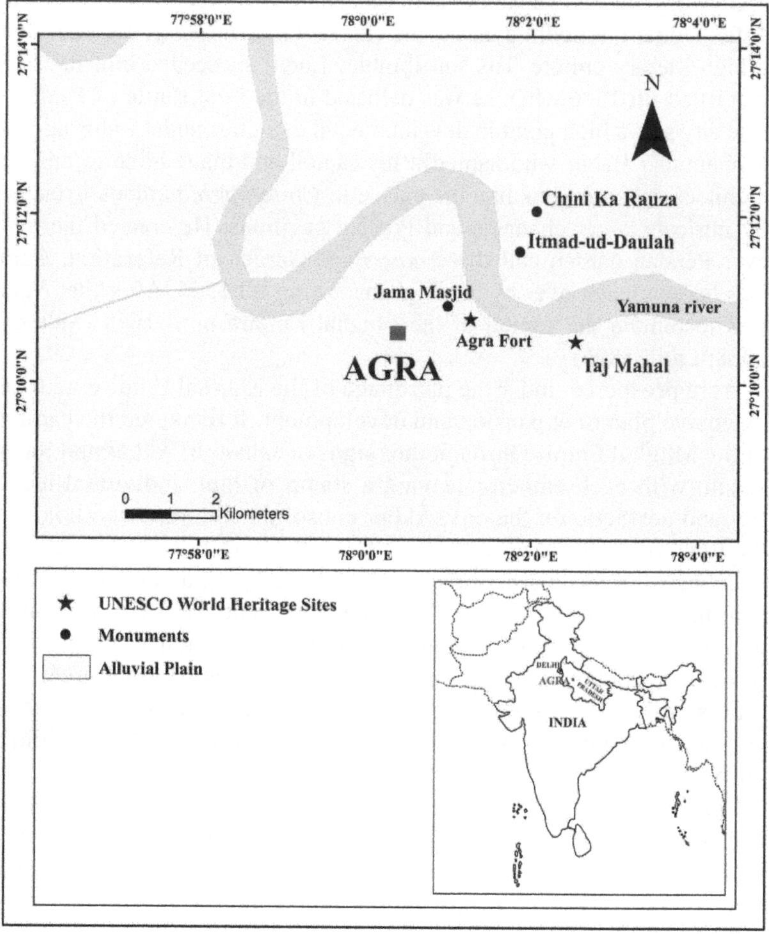

Figure 2.4 Map of Agra with locations of important monuments

2.3.1 *Chini Ka Rauza*

Built in 1635, Chini Ka Rauza is a mausoleum containing the Tomb of Allama Afzal Khan Mullah, the Prime Minister of Shah Jahan who was also a distinguished scholar and poet. The title "chini" refers to the glazed tile look of the monument, which has faded over a period of time. The tomb, with a unique architectural design and disproportionate dome is built facing Mecca (Archaeological Survey of India, Agra Circle, 2003).

2.3.2 *Itimad-ud-Daulah*

Known as "Baby Taj," this Mughal Mausoleum, located on the eastern bank of the Yamuna, is referred to as the draft of the Taj Mahal. Built between 1622 and 1628, it is a garden tomb replete with several ancillary buildings and structures as was the convention. The building was commissioned by Nur Jahan, the wife of Jahangir, for her father Mirza Ghiyas Begh, who had been given the title of *Itimad-ud-Daulah* (pillar of the state), and his wife Asmat Begum. It indicates the shift in Mughal architecture, from red sandstone based architecture to more and more refined, marble based structures with fine inlay work. It boasts very fine specimens of *pietra dura* work. The main mausoleum, set in a garden, is flanked by hexagonal pillars on all sides. Built of marble from Rajasthan, the structure is encrusted with semi-precious stones like lapis lazuli, jasper, topaz, onyx etc. (Archaeological Survey of India, Agra Circle, 2003; https://ia802807.us.archive.org/7/items/GLIMPSESOFAGRAMONUMENTS/GLIMPSES%20OF%20AGRA%20MONUMENTS.pdf).

2.3.3 *The Jama Masjid*

Dedicated to his eldest daughter Jahanaara Begum, the Jama Masjid or Friday Mosque at Agra was built in 1648 by Shah Jahan. It showcases extensive use of red sandstone and marble with bright blue walls. The central hall is accessed through a flight of 35 steps, making the structure very imposing. It highlights the mosque feature developed during Shah Jahan's reign: deepening of the central *iwan* (Koch 1991; https://ia802807.us.archive.org/7/items/GLIMPSESOFAGRAMONUMENTS/GLIMPSES%20OF%20AGRA%20MONUMENTS.pdf).

Akbarabad remained the capital under Aurangzeb till he moved it to Deccan in 1653. With the decline of the Mughals, the city came under the Marathas who held sway till 1803 when the British took over. The presidency of Agra was established in 1835 and suffered a famine from 1837–38. During the 1857 rebellion, two Infantry battalions of the British Army rebelled in Agra and marched to Delhi (Siddiqi, 2008). The position of the British was severely threatened with companies of Indian soldiers rebelling in Gwalior as well. This led to the British Army seeking refuge in the Agra Fort. As the rebelling soldiers moved to Delhi, the British were able to restore order and regain control of the city.

Thus one can see a whole trajectory of history as it played out in Agra: from its description in Ancient Indian epics to the stamp of Mughal monuments of the superlative order, to the fissures of Colonial intervention that introduced a new vocabulary, to a town with a distinct character in contemporary times.

Chapter 3

Repository of stones used in Delhi and Agra UNESCO Sites

Aravalli Mountain Belt and Vindhyan Basin

"What are men to rocks and mountains?"

Jane Austen

3.1 Introduction

This quote by Jane Austen emphasizes that in the light of the magnanimity of the Mother Earth and her resources, the human existence pales into insignificance. On the other hand, the human intervention has added a unique and artistic dimension to rocks in form of magnificent, tall stone edifices. The architectural heritage bears robust testimony to the cultural evolution of human society with time. The exemplary journey from natural rock shelters to the marble wonder Taj Mahal offers an abundant proof of the gradual increase in the human aesthetic evolution and abilities.

The excavations at the Purana Qila (Old Fort) narrate an impressive account on the occupancy of Delhi since prehistoric times. The tools from the Stone Age, relics of an ancient city '*Indraprastha*' (a strategic city mentioned in Hindu Epic Mahabharata), artefacts from Gupta and Mauryan periods, elaborate the existence of Delhi since the prehistoric time (www.livemint.com/Leisure/f2XeUnRMnc1mIlpVZDrZ0I/AravallisA-geological-marvel.html; www.thehindu.com/news/cities/Delhi/JNUs-rocking-pre-historic-legacy/article13377963.ece; www.thehindu.com/news/cities/Delhi/mahabharat-sites-continue-to-have-the-same-names-eventoday-b-b-lal/article5776270.ece; www.hindustantimes.com/delhi-news/the-cities-of-delhi-from-the-legend-ofindraprastha-to-qila-rai-pithora/story-B9mKCh192j5aVEcBUzJnYI.html; www.thehindu.com/news/cities/Delhi/the-discovery-of-indraprastha/article5772895.ece). Delhi has been the seat of power for centuries. The post 11th century accounts of Delhi reveal the existence of nine cities (Fig. 2.1 and 2.2; Chopra, 1976). For details on the contemporary 10th avatar of Delhi and the prior nine cities of Delhi from 11th to 20th century, the reader can refer to Chapter 2 of this book. An impressive and varied stone built architectural and monumental legacy projects Delhi as a tolerant land of diverse cultures and traditions. The modern day Delhi that evolved over a long period of time has grown into a truly cosmopolitan city with flavors of different cultures and traditions.

Agra, known for the Taj Mahal, easily one of the most admired monuments in the world, is not far from Delhi. The archaeological remains of the city of Agra, just like those of Delhi, hint at the habitation of the city prior to the Mughal period (Frowde, 1908). Chapter 2 of this book gives a brief account of the historical city of Agra. Agra and its vicinity display remarkable stone built architectural heritage during the Mughal period, it being the seat of Mughal dynasty. Agra acquired prominence during the reign of Emperor Akbar and Shah Jahan from the 16th to 17th century.

Delhi and Agra are situated on the banks of the river Jumna, a prominent river of northern India, which originates from the glacier Yamnotri in the Higher Himalaya. In consonance with the original Sanskrit name, the river is now re-designated as the Yamuna. The river has shrunk in size and also shifted its course with the passage of time. These two cities were built, destroyed and rebuilt several times and each time a distinctive character emerged. These two cities alone are witness to several fortified cities. Each city represents a unique period of history, narrating tales of devastation and reconstruction. The cities of Delhi and Agra boast of having unique stone built architectural sites that have been designated as World Heritage Cultural Sites by UNESCO (Fig. 3.1; https://whc.unesco.org/en/list/).

The building of each prior city of Delhi and Agra with forts, temples, mosques, palaces, mansions, *baolis* (stepwells), market places, sculptures etc. involved huge quantities of diverse building materials, most commonly, the stones/rocks. Thus, for obvious reasons stones/rocks used in the structures of Delhi and Agra were sourced majorly from the Aravalli Mountain Belt and the western sector of the Vindhyan Basin (Fig. 3.2).

Delhi and Agra are situated on the northern peripheral extension of the Aravalli Mountain Belt and the Vindhyan Ridge of Vindhyan Basin, respectively (Fig. 3.2). Owing to denudation for more than 1000 million years, past activities of quarrying (presently banned) and urbanization (Fig. 3.3), remnants of the Aravalli Mountain Belt are only sporadically exposed in parts of Delhi (www.livemint.com/Leisure/f2XeUnRMnc1mIlpVZDrZ0I/Aravallis-A-geological-marvel.html). The Vindhyan Basin rocks are not exposed on the surface in Agra city but are present beneath the alluvium (Verma, 1991; Ray *et al.*, 2003). Nevertheless, the Vindhyan rocks crop out as isolated ridges a few kilometers west-southwest of Agra (Fig. 3.2).

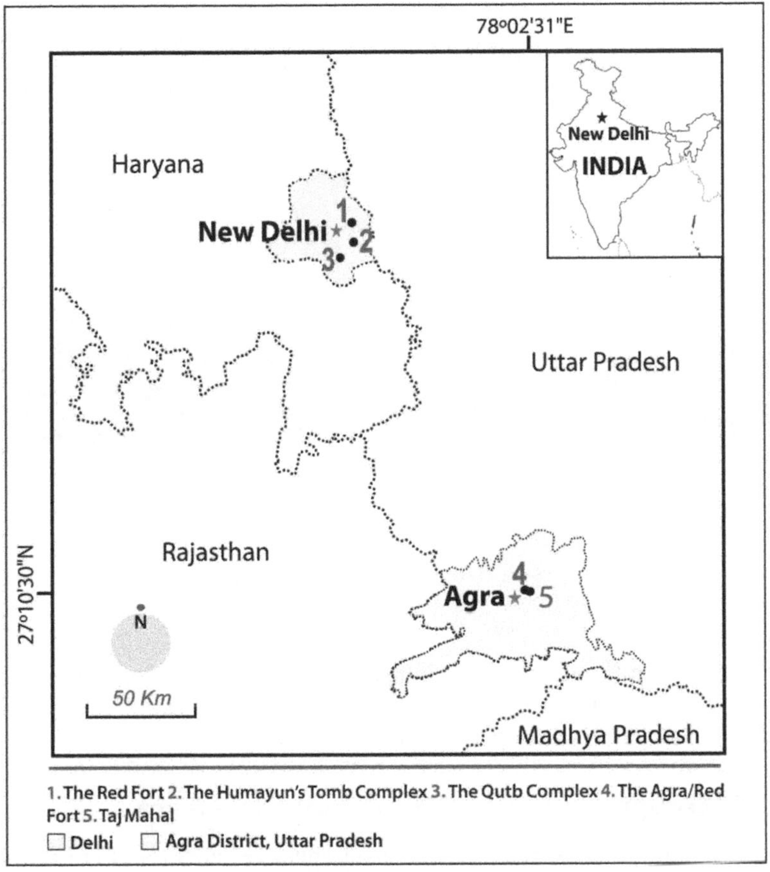

Figure 3.1 Map with location of UNESCO World Heritage Sites of Delhi and Agra

The current chapter focuses on the building stones involved in the construction of the five UNESCO designated World Heritage Sites from Delhi and Agra. For a succinct historical and architectural account on these monuments the reader may refer to Chapter 4 of this book. The building stones for the following World Heritage Sites of Delhi and Agra are discussed in this chapter:

1 Qutb Minar and its Monuments, Delhi (https://whc.unesco.org/en/list/233)
2 Humayun's Tomb, Delhi (https://whc.unesco.org/en/list/232)

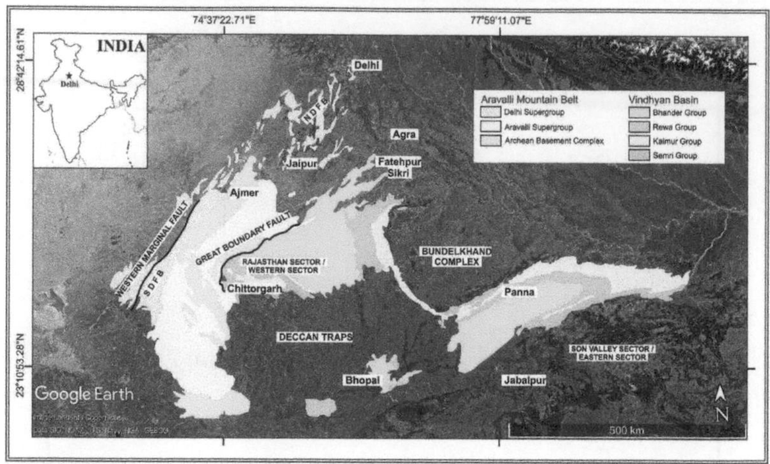

Figure 3.2 Geological map showing location of Aravalli Mountain Belt and Vindhyan Basin

(Source: after Prasad, 1984; Deb *et al.*, 2001; Cavallo and Pandit, 2008; Malone *et al.*, 2008; Base map from Google Earth)

3 Red Fort Complex, Delhi (https://whc.unesco.org/en/list/231)
4 Agra Fort, Agra (https://whc.unesco.org/en/list/251)
5 Taj Mahal, Agra (https://whc.unesco.org/en/list/252)

3.2 Geographic location of Delhi and Agra

Delhi, now also known as the National Capital Territory (NCT) of Delhi wide article 239AA and the 69th constitutional amendment, is a city and Union Territory of India. It is situated in the northern part of India bordered by the states of Haryana and Uttar Pradesh, covering an area of approximately 1484 sq km (Fig. 3.3). New Delhi was the new phase of Delhi developed between 1911 and 1940 when it was declared the capital of British India in 1911. New Delhi is a newly built city in contrast to the old city of Delhi situated on the banks of the river Yamuna. The peripheral limits of Delhi have expanded since the times of the first documented city of Lal Kot which existed during the 11th century (Fig. 2.3). The initial habitation concentrated mostly in south Delhi and gradually spread to the Delhi triangle (area between the Delhi ridge and

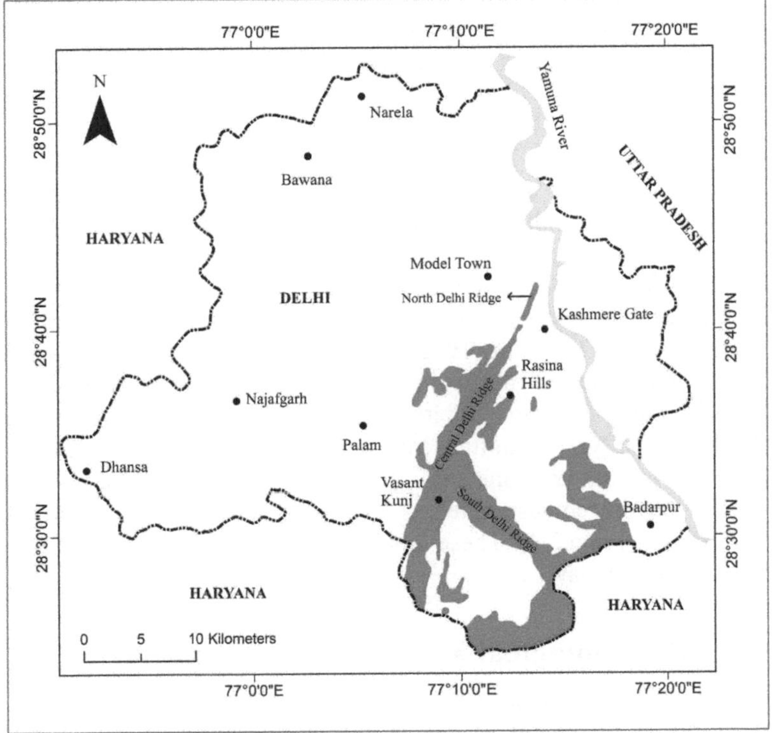

Figure 3.3 Map of present day Delhi
(Source: modified after Thussu, 2006)

the river Yamuna; Fig. 2.3 and 3.3) during the evolution of nine subsequent cities (Fig. 2.2 and 2.3; Chapter 2 of this book). Modern Delhi is an expanded city in terms of area and population both, spread far off from the Delhi triangle and often referred to as Delhi and National Capital Region (NCR) which includes Gurgaon, Noida, Faridabad, Sonipat and Bahadurgarh cities from the bordering states of Haryana and Uttar Pradesh (Fig. 3.3). Delhi occupies a strategic location on the map of India and is commonly referred to as "*dil*" of India which means "heart," not only because it hosts New Delhi, the capital city of India, but also because of its exquisite monumental heritage. Ever since India gained independence in the year 1947 the city has maintained its unique character built over centuries. This mesmerizing city with glimpses of the past/former cities and diverse stone built architectural heritage draws tourists from around the globe (www.thestatesman.com/features/for-delhi-the-heartof-india-1480115730.html).

Agra, just like Delhi, is also a historical city of India. It is part of Agra district of Uttar Pradesh state situated south of Delhi (Fig. 3.1). Agra city is known around the globe for the monument Taj Mahal to the extent that India has become synonymous with the Taj Mahal in the world's imagination. Agra was an important city of India during the Mughal rule in India and served as the capital city of the Mughal Empire for almost a century from 1556 to 1648 (Blochmann and Phillott, 1927; Nath, 1972; Habib, 1982), which is considered as the golden period of Mughal India and is fondly referred to by historians and art lovers for its days of opulence and magnificence (Nath, 1972). Agra is dotted with numerous heritage buildings and monuments built in stone during the last few centuries, and some of them have been designated as UNESCO World Heritage Cultural Sites (Fig. 3.1 and 2.4). These monuments are thronged by thousands of tourists each day.

3.3 Brief geological account of Delhi and Agra

Delhi holds a distinguished stature in terms of its geological attributes. The rocks of the Proterozoic age are overlain by the Quaternary alluvium/aeolian deposits (Fig. 3.4). Presently, almost 80% of the Delhi city is covered with sediments of Quaternary age divided into the Older and Newer alluvium with inselbergs of quartzite (Chopra, 1976; Thussu, 2006). The Delhi quartzite ridge – a part of the Aravalli Mountain Belt – is also commonly referred to as the "Delhi Ridge"/"Aravalli Ridge." It is broadly divided into the North, Central and South Ridge vis-á-vis its geographic location in Delhi (Fig. 3.4; Thussu, 2006; www.livemint.com/Leisure/f2XeUnRMnc1mIlpVZDrZ0I/Aravallis-A-geological-marvel.html). The sporadic quartzite ridges are cross cut by the Neoproterozoic post-Delhi intrusive phases represented by pegmatites and quartz veins (Table 3.1; Thussu, 2006).

The city of Agra is mostly occupied by the alluvium concealing all the older rocks present in the sub-surface. To the west and southwest of Agra city, the sporadic sandstone ridges, surrounded by the alluvial cover, belong to the Vindhyan Supergroup (Fig. 3.5; Frowde, 1908; Verma, 1991; Ray *et al.*, 2003).

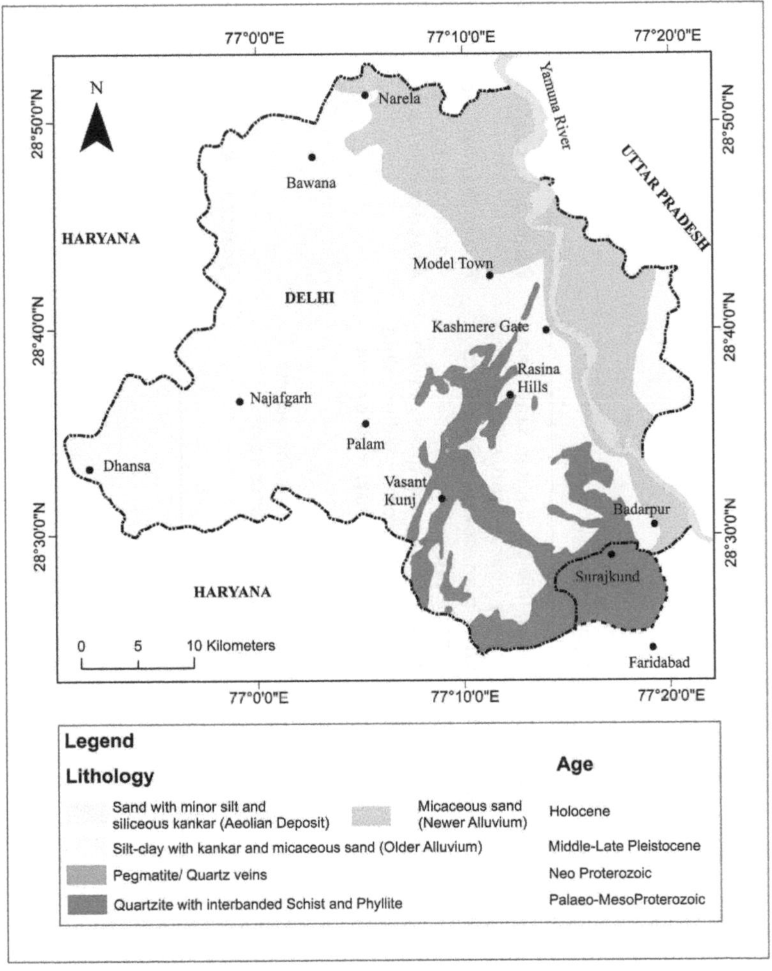

Figure 3.4 Geological map of Delhi

(Source: modified after Thussu, 2006)

Table 3.1 Stratigraphic succession of rocks in Delhi (Thussu, 2006)

Holocene	Yamuna Channel alluvium	Grey, fine to medium sand, grit with coarse sand, silt and clay	Point bars, channel deposits
	Yamuna Older Flood Plain and Terraces	Grey sand, coarse grit, pebble beds and minor clay	Palaeochannels, abandoned channels, meander scrolls, oxbow lakes
	Older Alluvium	Sequence of sand-silt-clay with yellowish brown medium sand with silt, kankar, with brown aeolian sand	Abandoned channels, meander scrolls
No sedimentation			
Neoproterozoic	Post Delhi Intrusive	Pegmatite, tourmaline-quartz veins and quartz veins	
Mesoproterozoic	Delhi Supergroup	Alwar Group	Quartzite with minor schist, tuff and ash beds

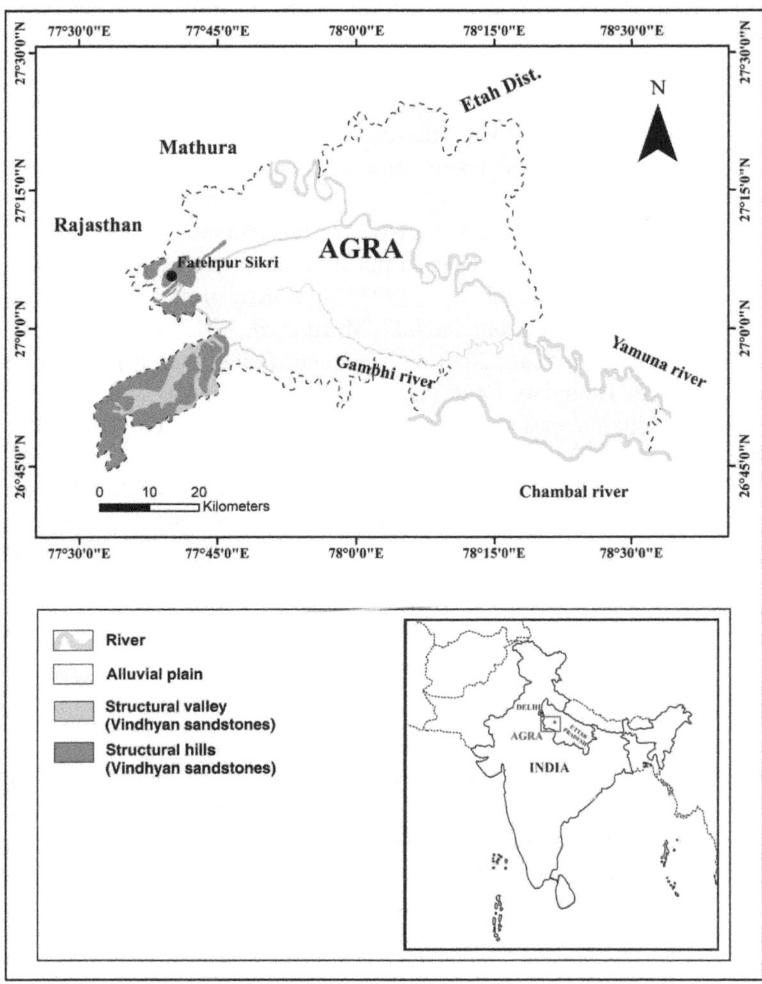

Figure 3.5 Geological map of Agra district

(Source: modified after Misra and Mishra, 2007)

3.4 UNESCO World Heritage Sites of Delhi and Agra vis-á-vis Aravalli Mountain Belt (AMB) and Vindhyan Basin

The UNESCO World Heritage Sites of Delhi and Agra, namely, Qutb Complex, Humayun's Tomb Complex, Red Fort Complex, Taj Mahal and Agra/Red Fort Complex are made in a variety of rocks/stones belonging to the Aravalli Mountain Belt (AMB)/Aravalli Orogen and the adjoining Vindhyan Basin (Fig. 3.2). The Precambrian age marble, sandstone and quartzite are the three most important rocks which have been used in the UNESCO designated architectural heritage of Delhi and Agra. It is, therefore, imperative to understand the Precambrian geological framework of the AMB and the adjoining Vindhyan Basin.

The regional geological framework of the Aravalli Orogen and the Vindhyan Basin has been worked out by numerous geologists (Gupta, 1934; Heron, 1953; Gupta *et al.*, 1997; Sinha-Roy *et al.*, 1998; Roy and Kataria, 1999; Roy and Jakhar, 2002; Meert *et al.*, 2013; Mckenzie *et al.*, 2013; Gilleaudeau *et al.*, 2018 and references cited therein). The intervening Great Boundary Fault (GBF) between the AMB and Vindhyan Basin delimits the eastern and western extremities of the AMB and the Vindhyan Basin, respectively (Fig. 3.2).

The Aravalli Mountain Belt extending in NNE-SSW direction for over 750 kms, between Delhi and Gujarat, forms a prominent geomorphic entity of northwestern India (Fig. 3.6). The Precambrian rocks of AMB have been divided into the following three stratigraphic units:

3 Delhi Supergroup
2 Aravalli Supergroup
1 Banded Gneissic Complex

The Banded Gneissic Complex (BGC) is one of the oldest (Archean age) cratonic nuclei in the Indian Peninsular Shield around which crustal growth took place in successive stages. The Aravalli Supergroup rocks are exposed in the southern and southeastern parts of the AMB, whereas the Delhi Supergroup rocks are exposed in the southwestern, central and northeastern parts of the AMB (Fig. 3.6).

The BGC comprises Tonalite-Trondhjemite-Granodiorite (TTG), migmatitic gneisses, amphibolites, schists, quartzites, calc-silicate rocks, granulites, granites and pegmatites. The BGC is also referred to as the Mewar Gneiss and the Bhilwara Supergroup (Wiedenbeck *et al.*, 1996; Roy and Kröner, 1996). The BGC is overlain by the Paleoproterozoic Aravalli, Meso-Neoproterozoic Delhi and Vindhyan Supergroups (Rasmussen *et al.*, 2002; Ray *et al.*, 2002; Gregory *et al.*, 2006; Mckenzie *et al.*, 2013 and references cited therein). The Aravalli Supergroup primarily consists of metavolcanics, with low to medium

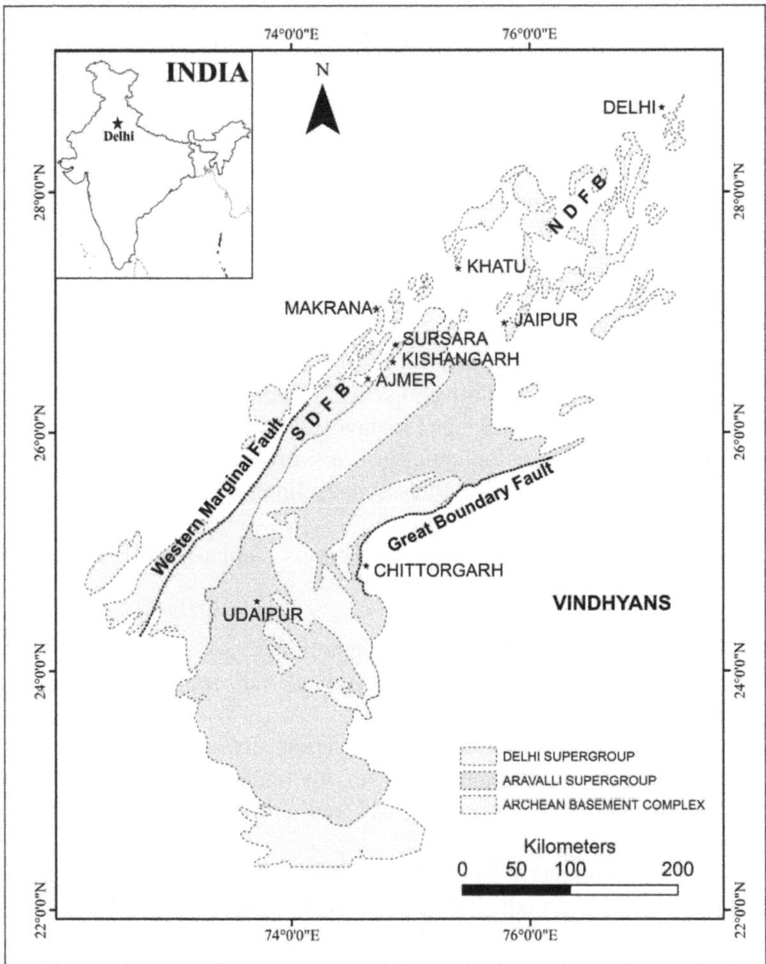

Figure 3.6 Geological map of Aravalli Mountain Belt

(Sources: Deb *et al.*, 2001; Cavallo and Pandit, 2008)

grade metasediments comprising quartzites, phyllites, schists, stromatolitic dolostone, phosphorite and greywacke (Heron, 1953; Roy and Paliwal, 1981; Gupta *et al.*, 1981, 1997; Sinha-Roy *et al.*, 1998; Roy and Kataria, 1999). The Aravalli Supergroup rocks are intruded by mafic and ultramafic rocks, besides the post-Aravalli and post-Delhi granites and pegmatites (Sharma, 1953; Srivastava, 1988). The Delhi Supergroup extends for a distance of ~700 km from Delhi and south Haryana to Idar in north Gujarat, as a narrow, linear belt that governs the NNE-SSW orographic trend of AMB. The Delhi Supergroup is divided into the older North Delhi Fold Belt (NDFB) and a younger South Delhi Fold Belt (SDFB) (Fig. 3.6; Sinha-Roy, 1984; Deb *et al.*, 2001; Saha and Mazumder, 2012; Mckenzie *et al.*, 2013 and references cited therein).

The NDFB is divisible into two units (Heron, 1917a, 1917b; Heron, 1953; Gupta *et al.*, 1981, 1997; Singh, 1984a, 1984b; Singh, 1988; Sinha-Roy *et al.*, 1998; Mckenzie *et al.*, 2013 and references cited therein):

2 Ajabgarh Group: phyllite and schist, marble, minor quartzite, calc-schist, calc-gneiss and mafic metavolcanics
1 Alwar Group: conglomerate, grit, arkose, feldspathic and ferruginous quartzite, minor phyllites, schists, limestone and metavolcanics

The SDFB is also divided into two Groups (Sinha-Roy, 1984; Gupta *et al.*, 1997):

2 Kumbhalgarh Group: pelitic schist, pelitic gneiss, quartzite, phyllite, greywacke, pure/impure marbles and calc-silicate rocks (=Ajabgarh Group)
1 Gogunda Group: metamorphosed arenaceous sediments, mainly conglomerates and quartzite with intercalated phyllites (=Alwar Group; Saha and Mazumder, 2012 and references cited therein)

The Raialo Group has a debatable position in the stratigraphic sequence of the Aravalli Mountain Belt. It has been renamed as the Rayanhalla Group and is considered part of North Delhi Fold Belt (Roy, 2006; Saha and Mazumder, 2012). It comprises dolomitic marble, metavolcanics and quartzites (Heron, 1953; Roy, 2006; Saha and Mazumder, 2012 and references cited therein).

The Vindhyan Supergroup rocks are exposed in the southeastern part of the state of Rajasthan and adjoining parts of Uttar Pradesh. These form the westernmost confine of the Vindhyan Basin that extends in E-W direction up to north-central India covering the states of Uttar Pradesh and Madhya Pradesh (Fig. 3.7). It is believed to have evolved as an intra-cratonic rift basin or as a peripheral foreland

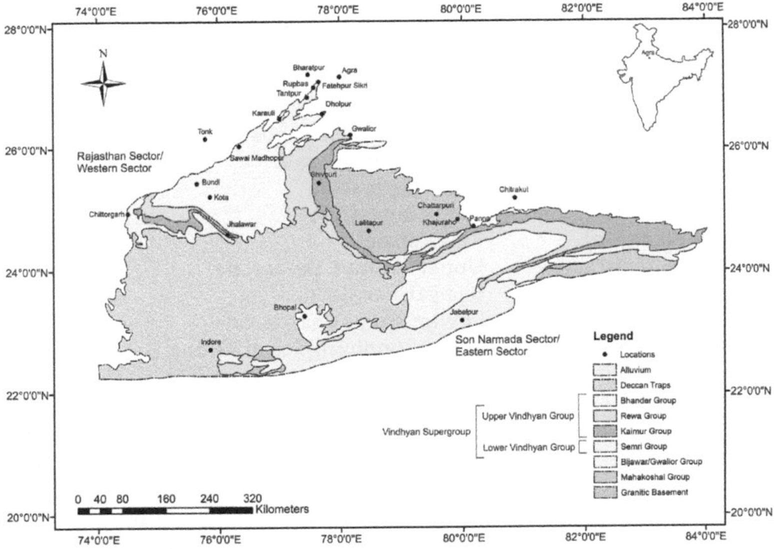

Figure 3.7 Geological map of Vindhyan Basin

(Source: adapted from Prasad, 1984; Malone *et al.*, 2008)

basin (Fig. 3.7; Ray *et al.*, 2003 and references cited therein). It is constrained in the west by the GBF, in the south and southwest by the Deccan Traps, in the southeast by the Bijawar and Mahakoshal groups and in the north by the basement rocks of Bundelkhand Craton and the Gangetic alluvium (Fig. 3.7). With reference to the Bundelkhand Craton and the Deccan traps, the Vindhyan Basin is geographically divided into the western Vindhyan sector known as Rajasthan sector and the eastern Vindhyan sector known as Son valley sector (Fig. 3.7; Prasad, 1984; Ray *et al.*, 2003; Mckenzie *et al.*, 2013; Gilleaudeau *et al.*, 2018 and references cited therein). The Vindhyan Basin represents a huge basin, with an extensive aerial spread roughly covering an area over 100,000 sq km (Bose *et al.*, 2001; Gilleaudeau *et al.*, 2018). It primarily comprises almost undeformed and unmetamorphosed sedimentary rocks. At places, the Vindhyan Supergroup strata attain a maximum thickness of about 4.5 km (Gilleaudeau *et al.*, 2018). The subdivisions of the Vindhyan

Table 3.2 Generalized lithostratigraphy of Vindhyan Supergroup

Group	*(After Bose* et al., *2001)*	
Bhander (1000 m) (Upper Vindhyan)	Upper Bhander sandstone	
	Sirbu shale	
	Lower Bhander sandstone	
	Bhander limestone	
	Ganurgarh shale	
Rewa (2000 m) (Upper Vindhyan)	Rewa sandstone	
	Rewa shale	
Kaimur (400 m) (Upper Vindhyan)	Upper Kaimur sandstone	
	Bijaigarh shale	
	Lower Kaimur sandstone	
Semri (1300 m) (Lower Vindhyan)	Lower Vindhyan Semri Group	
	Formation	Member
	Rohtas	Rohtas limestone
		Rampur shale
	Kheinjua	Chorhat sandstone
		Koldaha shale
	Porcellanite	Porcellanite
	Kajrahat	Kajrahat limestone
		Arangi shale
	Basement rocks	Bundelkhand granite

(Source: Majid *et al.*, 2012)

Supergroup rocks are furnished in Table 3.2. The Lower Vindhyan and Upper Vindhyan are separated by a major angular unconformity (Chakraborti *et al.*, 2010; Saha and Mazumder, 2012). The Lower Semri Group is dominantly composed of arenaceous (sandstone), argillaceous (shale) and calcareous rocks. The upper Kaimur and Rewa groups dominantly comprise siliciclastic rocks in contrast to the Bhander Group which comprises arenaceous (sandstone) and carbonate rocks (Gilleaudeau *et al.*, 2018 and references cited therein).

3.4.1 *Makrana Marble of Delhi Supergroup*

The Makrana marble bands are exposed in the vicinity of Makrana town in Nagaur district of Rajasthan, which is situated at about 130 km WNW of Jaipur (capital city of Rajasthan; Fig. 3.6). The

Table 3.3 Stratigraphic succession around Makrana, Nagaur District (Bhadra *et al.*, 2007)

Quaternary				Aeolian mobile sand with calcareous clay or silt with polymictic conglomerate and grit
Upper Proterozoic	Erinpura Igneous Suite			Biotite granite, Pegmatite, Amphibolite
Lower to Middle Proterozoic	Delhi Supergroup	Punagarh Group	Bombolai Formation	Phyllite, Impure limestone, calc silicate rock
		Kumbhalgarh/ Ajabgarh Group	Ras/Ajmer Formation	Makrana marble and dolomitic limestone/ quartzite

Makrana marble belongs to the Ras Formation of Kumbhalgarh Group of SDFB (Table 3.3; Fig. 3.6). The Makrana marble bands, trending NNE-SSW, are exposed to the west of Makrana town (Fig. 3.8). The prominent five marble bands are locally named after the hill ranges belonging to SDFB: (1) Devi-Gunawati Range, (2) Dungri Range, (3) Pink Range, (4) Makrana Kumhari Range and (5) Borawar Kumhari Range (Fig. 3.8; Paliwal *et al.*, 1977; Natani, 2000, 2002; Natani and Raghav, 2003). Each band is a combined unit of massive marble with intercalated calc-silicate and/or calcareous quartzite. The bands, confined between Matabhar in the north and Mored in the south, collectively are about 13 km in length and 1.6 km in average width (Fig. 3.8; Natani, 2000, 2002; Natani and Raghav, 2003). The marble bands are isoclinally folded and covered by the Quaternary aeolian deposits majorly comprising calcareous sand and grit (Fig. 3.8; Natani, 2000, 2002; Natani and Raghav, 2003; Bhadra *et al.*, 2007).

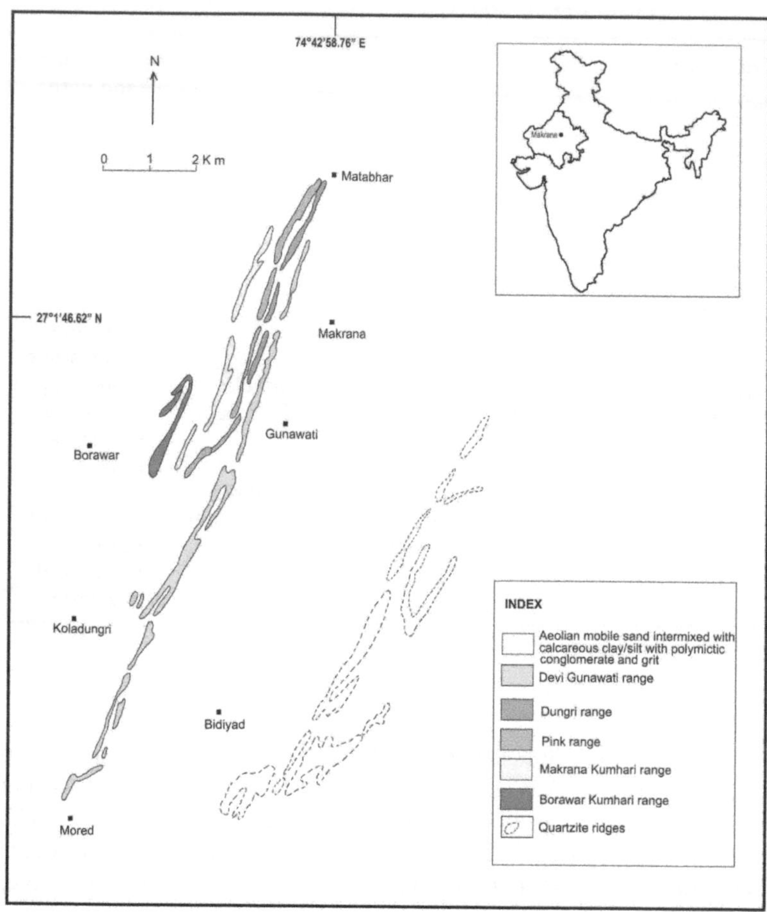

Figure 3.8 Geological map of area around Makrana town marked with bands of Makrana marble, Nagaur district, Rajasthan

(Source: modified after Natani and Raghav, 2003)

3.4.2 *Bhander sandstone of Western Sector Vindhyan Supergroup*

The Bhander Group, youngest stratigraphic unit of Upper Vindhyan Supergroup, overlies the Rewa Group with a minor disconformity in between. The Bhander Group is further divided into arenaceous, argillaceous and calcareous members viz., Ganurgarh shale, Bhander limestone, Lower Bhander sandstone, Sirbu shale and Upper Bhander sandstone members (Table 3.2). It comprises sandstones with minor shale and limestone. The Upper Bhander sandstone member is well exposed in Fatehpur Sikri, Rupbas, Bansi Paharpur, Karauli, Baretha, Rudawal, Bundi, Kota, Bharatpur, Sawai Madhopur, Tonk, Jhalawar, Karauli, Bhilwara and Chittorgarh districts in the Western Sector of Vindhyan Supergroup (Fig. 3.7; Sarkar *et al.*, 2004; Majid *et al.*, 2012; Kaur *et al.*, 2019b). The Bhander sandstone is medium to fine grained, compact and displays hues of red and yellow and streaks as a result of minor lithological contrast and cement. The Bhander sandstone exhibits parallel laminae, cross beds and ripple marks, thus imparting varied and vibrant textures to the thickly bedded sandstones (Fig. 3.9a–b). The Bhander Group sandstone has been commonly used in several stone built architectural structures in Delhi and Agra.

3.4.3 Delhi Quartzite of the Delhi Supergroup

The topography of Delhi is controlled mainly by the Alwar Group quartzites exposed in the Delhi ridge, small hills and plateau along with the Quaternary alluvium. The Delhi Ridge primarily composed of Delhi quartzites represents the northernmost extremity of the AMB. The Quaternary aeolian sand and fluvial alluvium veneer covers almost 80% of Delhi (Fig. 3.4; Table 3.1; Thussu, 2006). The Delhi Ridge which was continuous for kilometers about a century ago is now sporadically exposed owing to rapid urbanization in Delhi in the last 50 years. The general trend of the Delhi ridge is NNE-SSW in the northern and central parts of the city and it swings to WNW-ESE or E-W in the south central and south Delhi (Fig. 3.4). The Delhi Ridge gets broader in the southern parts of the city and the adjoining Surajkund (Faridabad), Haryana. The south Delhi ridges are broader and merge with the plateau south of Tughlaqabad (Fig. 3.4). The quartzites contain recrystallized quartz, thereby making them inherently hard, compact and resistant to chemical weathering.

The Alwar Group quartzites exposed in the Delhi Ridge are milky white to grey and locally pinkish and reddish in color (Fig. 3.10a and b). At places these are streaked with reddish spots due to the presence of iron oxide and pyrite grains. The quartzites exhibit sub conchoidal fracture and sharp edges when hammered. Ripple marks and current bedding are commonly observed. The quartzites mostly contain two prominent sets of joints (Chopra, 1976; Thussu, 2006).

Figure 3.9 Exposure of Bhander Group Sandstone in (a) Rupbas and (b) Bansi Paharpur

(Photos: Parminder Kaur)

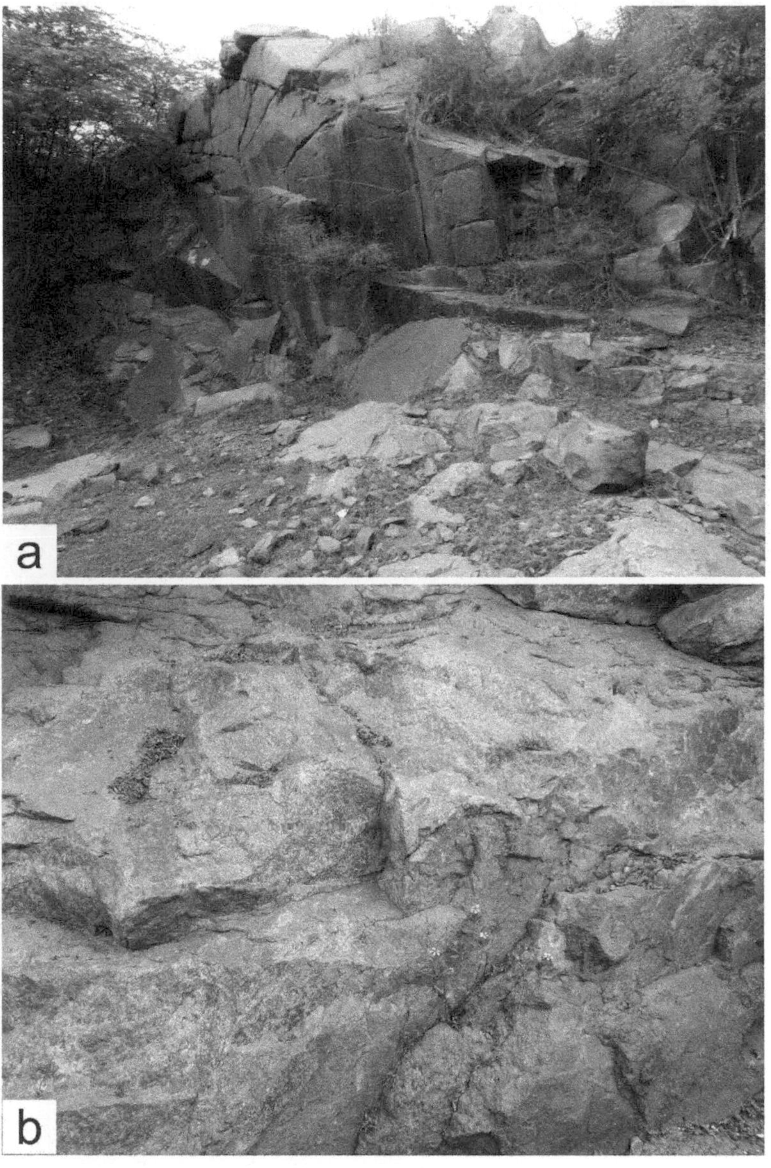

Figure 3.10 Exposure of Delhi Quartzite in (a) Lal Kuan and (b) American Embassy School in Delhi

(Photos: Gurmeet Kaur)

3.4.4 Petrography and mineralogy of Makrana Marble

The Makrana marble quarries produce a flawless white variety along with other varieties, which have hues of grey, pink and brown or well defined bands of grey, greyish black and greenish grey colors imparting a patterned texture to the marble (Fig. 3.11a and b). The marble varies from fine to coarse-grained varieties. These exhibit granoblastic texture formed by interlocked recrystallized calcite grains, which imparts on

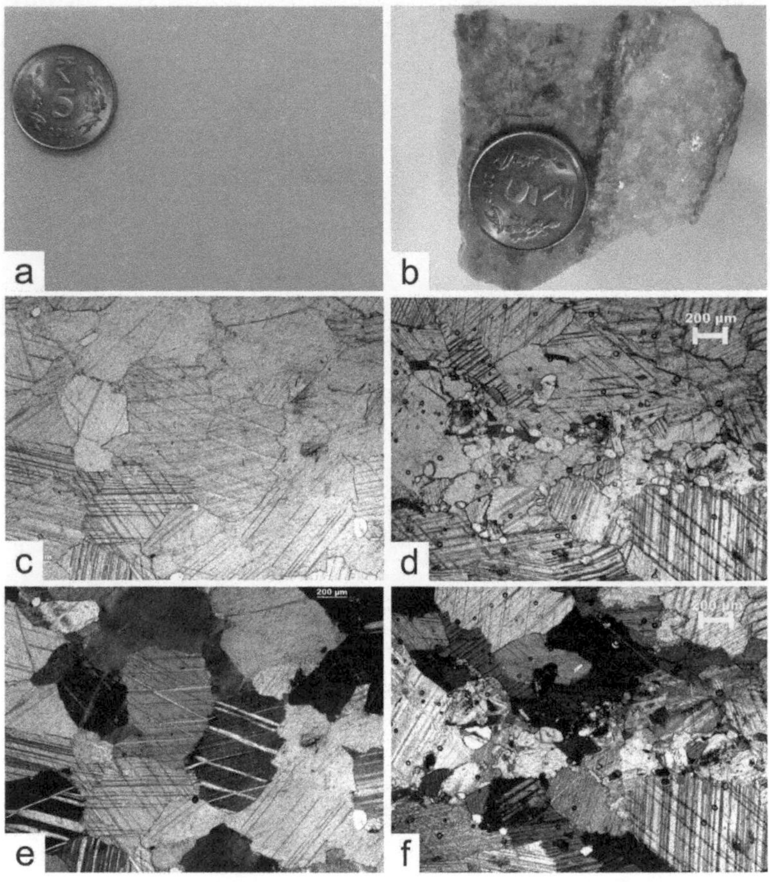

Figure 3.11 (a and b) Hand specimen of Makrana white marble and Albeta marble; photomicrographs of Makrana white marble (c and e) and Albeta marble (d and f) in Plane Polarized Light (PPL) and Cross Nicols (XN), respectively

the marble a high bearing strength. The Makrana white is pure calcitic marble with up to ~100% calcite crystals with no other major or minor mineral and thus can be termed as monomineralic (Fig. 3.11c and e). The banded brown, grey/green and pink varieties contain quartz, biotite, diopside, tremolite, actinolite, olivine and sometimes serpentine in traces. The rock shows mosaic texture in hand specimen and granoblastic texture in thin section (Garg *et al.*, 2019; Fig. 3.11c–f). Makrana white marble with greenish grey bands display granulose texture comprising calcite, talc, wollastonite, forsteritic olivine and serpentine (Garg *et al.*, 2019; Fig. 3.11d–f). The streaks, lenses and bands of grey and brown in white variety of Makrana marble can be attributed to the earlier-mentioned silicate mineral impurities.

3.4.5 Petrography and mineralogy of the Vindhyan Sandstone

Based on petrographic studies the Bhander Group sandstone, collected from Rajasthan sector, can be termed as "quartz arenite" as per classification of Folk (1980) and others (Bhardwaj, 1970; Bose and Chakraborty, 1994; Banerjee and Banerjee, 2010; Majid *et al.*, 2012; Khan *et al.*, 2013; Sen *et al.*, 2014; Verma and Shukla, 2015; Quasim *et al.*, 2017; Kaur *et al.*, 2019b and references therein). The petrographic studies of Vindhyan Sandstone varieties from Rupbas, and Bansi Paharpur (Fig. 3.12a–f), which were commonly used as building stone in most of the architectural heritage of Delhi and Agra, reveal that these are quartz arenite type. The sandstones are dominantly monominerallic with quartz as the major framework mineral grain, followed by rock fragments and feldspars, wherein matrix (<15%) consists of silt and clay. The quartz is colorless to dirty white, subangular to subrounded, well sorted and medium to fine grained (Fig. 3.12c–f). In general, cement is ferruginous in the red sandstone and dominantly siliceous with traces of ferruginous in beige sandstones. Petrographic examinations of sandstones of the earlier-mentioned localities indicate their high textural maturity and the monominerallic quartz composition, which accounts for their strength and durability (Kaur *et al.*, 2019b). Majid *et al.* (2012) have classified the Upper Bhander sandstone from the vicinity of Fatehpur Sikri as quartzarenite with quartz being the dominant mineral. Muscovite, biotite and mostly altered feldspars are also reported from these sandstones. Chert, phyllite and schist constitute the common rock fragments in rock. Heavy minerals in the Upper Bhander sandstones include ilmenite, magnetite, tourmaline, zircon, garnet and rutile (Majid *et al.*, 2012).

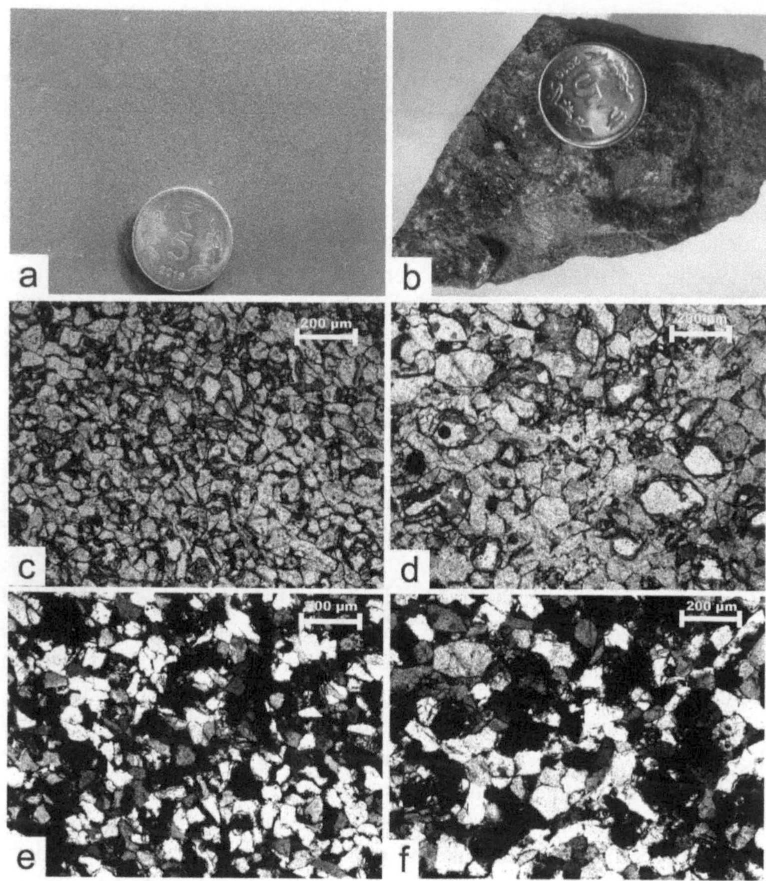

Figure 3.12 (a and b) Hand specimen of Vindhyan Sandstone from Rupbas and Bansi Paharpur; Photomicrographs of Rupbas (c and e) and Bansi Paharpur (d and f) in Plane Polarized Light (PPL) and Cross Nicols (XN), respectively

3.4.6 Petrography and mineralogy of the Delhi Quartzite

The Delhi quartzite exposed in south Delhi is predominantly made up of recrystallized quartz (Fig. 3.13a). The quartz grains are anhedral and have typical granoblastic polygonal texture showing triple points (Fig. 3.13b–d). Most of the quartz grains are non undulose forming a

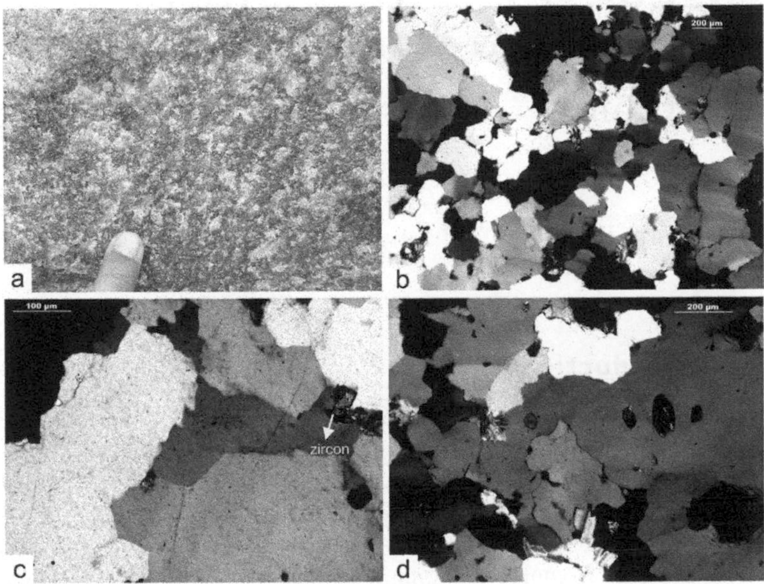

Figure 3.13 (a) Hand specimen of Delhi quartzite from Lal Quan; Photomicrographs of Delhi quartzite (b, c and d) in Cross Nicols (XN)

mosaic of the grains. Tourmaline, biotite, muscovite, zircon, pyrite and a few opaque minerals are sporadically observed in the quartzite samples (Fig. 3.13b–d; Chopra, 1976; Tripathi and Rajamani, 2003). Tripathi and Rajamani (2003) have reported abundant tourmaline in a few samples of the Delhi quartzites. In almost all the samples feldspar is typically rare or absent. Reddish brown coating along the grain boundaries and interstitial spaces of a few quartzites is due to presence of iron oxide. Appreciable amounts of opaques (perhaps pyrite and Fe-Ti oxides) are randomly seen in the thin sections of quartzites (Chopra, 1976; Thussu, 2006).

Chapter 4

UNESCO Heritage Sites of Delhi and Agra

An account

4.1 Introduction

The monuments shortlisted for this study encompass an interesting temporal and aesthetic expanse. They help us link the dots in Indian history from the 12th century to 17th century, indicating important events that not only concern the cities of Delhi and Agra (where these monuments are located) but illustrate crucial linkages in the history of the nation as such. The Qutb Complex of monuments indicates the first ever Islamic intervention in the architectural vocabulary of India, the first ever mosque on the Indian soil that still stands extant, to the first ever garden tomb (a rampant convention in Islamic worldview) in Humayun's Tomb, to the Agra Fort relating a virtual history of the great Mughals to the Red Fort that symbolized the high point of Mughal dominance, to the Taj Mahal that well nigh touched the pinnacle of Indo-Islamic refinement in architecture, horticulture and allied arts.

The UNESCO has identified these monuments (through a multi-pronged criteria) to be included in the World Heritage List as monuments that reflect an exemplary level of built heritage, historical-cultural and aesthetic value (http://whc.unesco.org/en/criteria/). The plans used are illustrative of the concerned monument complex. The significant monuments pertaining to each complex are identified and described. Not all monuments could be taken up for a detailed description due to the scope of the present work.

4.2 Qutb Minar and its Monuments, Delhi

In 1993, the Qutb Complex (Fig. 4.1) was inscribed in the World Heritage List by UNESCO under criteria (iv) that recognizes it "to be an outstanding example of a type of building, architectural or technological ensemble or landscape which illustrates significant stage(s) in human history" (https://whc.unesco.org/en/list/233). Further, through this inscription, a level of recognition of the importance of heritage, coupled with a dynamic flow of resources towards the upkeep and management of this cultural heritage is ensured.

Apart from the visually stunning Qutb Minar, which could easily be one of the most widely recognized monuments in the world, the sprawling Qutb Complex at Mahrauli includes the Quwwat-ul-Islam mosque, the Alai Darwaza, the tombs of Iltutmish and Alauddin Khilji and other

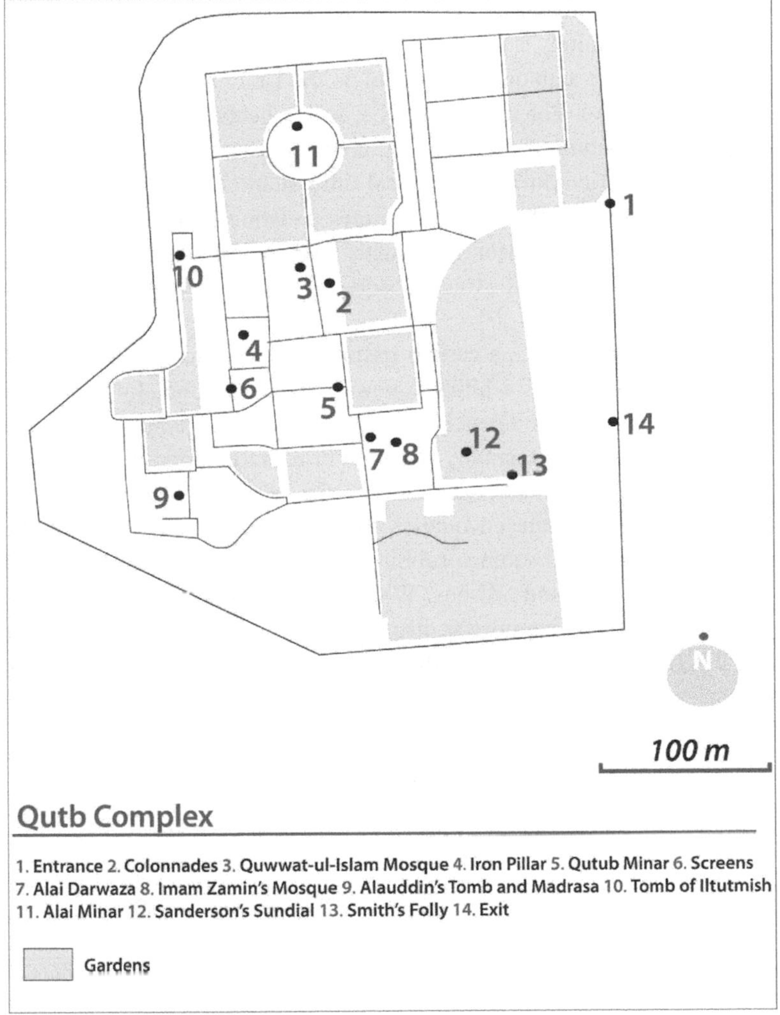

Figure 4.1 Plan of Qutb Complex

(Source: Traced from google platform 2019)

minor structures. The complex, enclosing monuments dating to 12th century to later dynasties, not only presents a visually stunning skyline but also acts as a visual reckoner of the range of influences that the city of Delhi has witnessed over time. Architecturally and politically, the Qutb Complex captures the myriad layers embedded in the expanse of the city. It is the story of how power makes inroads over time, and adapts itself to the existing climate of a place. The ineluctable dance of power and destruction is a singular fact visible in the precincts of the Qutb.

The Qutb definitely is a statement, even today, as the spire of the minaret is visible, like a ubiquitous symbol on the Delhi horizon for denizens and visitors alike. For the metropolis, it has become a visual emblem of sorts, a signature of the medieval connect, a totem of the many layers of Delhi buried through historical time, marked by a strident visual language of the minaret. It also connotes a distinct symbolism, the first brush of the Indian subcontinent with Islamic influence, the complex still enclosing the oldest extant mosque in India: the Quwwat-ul-Islam mosque.

The Qutb Complex is a certain pastiche of influences. While it does mark the beginning of a hitherto new influence on the skyline of the nation, in itself the monument became a field of experimentation: Islamic architecture has fundamentally relied on brick and mortar, whereas temple architecture had made use of stones for millennia (Mitra, 2002). In the Qutb Complex one encounters Persian architectural forms of the arch, calligraphy and squinch rubbing shoulders with the age old art of Hindu architecture and artisans. While some people see remnants of a violent past, it will be more fruitful to see it in the context of the long line of syncretic influences that the subcontinent was to witness for many centuries hereafter.

4.2.1 Qutb Minar

> One half of our existence is a blank;
> A mighty empire hath forgetfulness!
> History is but a page in the great past,
> So few amid time's records are unsealed.
> Here is a mighty tower: ere it was raised
> Its builders must have had wealth, power, and time,
> And a desire beyond the present hour.
>
> Letitia Elizabeth Ladon "The Qootub Minar" (1833)

The minaret was erected to mark the victory of Qutbuddin Aibak, who began its construction in the 12th century. A later *nagari* inscription refers to the Qutb as Alau-Din's *Vijaya-Stambh* (victory tower) (Sharma, 2015). Qutbuddin Aibak started the construction around 1192. There are two theories of its origin: that the minaret was named after Qutbuddin Aibak, or another that traces the origin to Qutbuddin Bakhtiar Kaki, the prominent Sufi saint. His successor Shamsuddin Iltutmish completed three more storeys by 1220. Qutb Minar (Fig. 4.1; point 5) is a 72.5 m high tower (minaret) of five storeys, with the use of red sandstone (varying colored hues), marble and quartzite stone (Fig. 4.2a). The inner part of the minar, from the basement to the fourth floor is of grey quartzite stone and farther upward is of red sandstone. The masonry of the minar is also worth describing – the innermost core was built with ashlar masonry of Delhi stone, the hearting was done using rubble stone masonry, and the outermost veneering was done with Agra sandstone (Balasubramaniam, 2005). Close inspection reveals calligraphy inscriptions from the holy Quran (Fig. 4.2b). Since Islam prohibits idol making and imagery of holy personages and entities, it has relied on calligraphy as a means to crystallize the divine word. The zeal to lend it a unique aesthetic over millennia has led to highly developed calligraphy art that touches a pinnacle in the later Mughal architecture. Since Hindu artisans and craftsmen were part of the building process, an architectural vocabulary

Figure 4.2 (a) The Qutb Minar; (b) calligraphy inscriptions from the Quran on the bands of the Minar

(Photos: Gurmeet Kaur)

evolved which is also referred to as Indo-Saracenic/Indo-Islamic architecture (Page, 1926).

4.2.2 Quwwat-ul-Islam mosque, screen and Iron Pillar

Qutb Minar is part of a larger complex that also includes the Quwwat-ul-Islam mosque (Fig. 4.1; point 3), which is considered to be the oldest extant mosque in India. Built in 1192 AD to mark the victory of Mohammad Ghuri, who invaded India, leading a force from Ghur in Afghanistan (Mitra, 2002). This fact is also recorded on the main eastern entrance. He installed his trusted lieutenant, Qutbuddin Aibak, who decided to mark this military victory with symbols of might. This is recorded on the gateways of the mosque erected in the north and the east. This is when the Quwwat-ul-Islam mosque was constructed over the captured citadel of Qila Rai Pithora. According to Spear (1997), Aibak did not have any skilled masons with him, so he had the workers of Prithviraj's Lal Kot build the mosque for him. It is architecturally, culturally and aesthetically a landmark because it introduced a new layer of influence in the Indian space that was to then dominate India for many successive centuries.

The mosque courtyard is surrounded by pillared cloisters and stepped entrances on the north, east and south sides. Corbelling technique was used to build the arches. It is an open rectangular courtyard, hemmed in by a colonnaded veranda running across one length of the structure (Fig. 4.3a). The veranda is made of temple pillars in the style of the Jain temples at Mount Abu. Spear (1997) makes a mention of a similar mosque built by Qutbuddin in Ajmer tellingly called *Arhai-din-ka-Jhompra* because it was built in two and a half days: perhaps a medieval equivalent of the prefab. There are secluded *zenana* quarters located on the mezzanine floor, accessible by a narrow discrete staircase. The running roof is supported by decorative columns, each one distinct in its style (Fig. 4.3b), so much so that no two columns are identical. These columns belong to the various temples that were destroyed by the invading armies.

The mosque was composed entirely of material taken from 27 Hindu and Jain temples destroyed by Ghur soldiers. These columns display a broad range of relief and iconography like sculpted lotus flowers, *kalasa*, flowering vines and such like typically Hindu motifs (Fig. 4.3c). To get a coherent structure fabricated from such a disparate range of techniques and material is in itself a test of ingenuity. The relief idols have been somewhat defaced but the remains reveal that it was a hurried job. But within the same complex of the mosque, advancement in Indo-Islamic architecture is visible, especially with the screen added by Qutbuddin

Figure 4.3 (a) Colonnaded verandah of Quwwat-ul-Islam mosque; (b) roof of the mosque supported by ornate pillars; (c) relief work and indigenous motifs on mosque pillars

(Photos: Gurmeet Kaur)

Aibak where the calligraphy shows a marked influence of the Hindu worldview, each stroke ending in the characteristic flower bud, an oft repeated motif in temple decoration (Mitra, 2002). Similarly, the Hindu masons built the first ever pointed arch in India here without the knowledge of keystone, therefore building arches with modifications. The exuberance of floral iconography in Aibak's time gradually gives way to geometric motifs in an Islamic fashion in Iltutmish's time. The mosque is in a state of neglect, but features like indigenous corbelled arches, floral motifs and geometric patterns adorn the surfaces and architecture (Fig. 4.3a–c). The core of the mosque is rubble masonry with cladding of red sandstone. This mosque remained the Jama Masjid of the Sultans of Delhi for 30 years (Spear, 1997). It was Iltutmish who decided to add another layer of grandeur to the mosque. He got three stately arches added to the original monument. Thirty years down the line, the level of expertise in handling Persian features had definitely improved and artisans were also brought from Ghur to work on the additions (Sharma, 2015). So the difference between the tentative vocabulary of the original structures of the mosque vis-à-vis the later additions is palpable.

A flamboyant red hue sandstone screen (Fig. 4.1; point 6) stands across the front of the mosque and attests to the gradual Islamization of Indian architecture, especially because the screen was a work in progress through the reigns of Qutbuddin, Iltutmish and Alauddin Khilji. It comprises of a central arch flanked by two smaller arches on either side (Fig. 4.4). It has calligraphic inscriptions in *Naskh* characters embedded in rich floral designs reminiscent of Hindu temple motifs. However by Iltutmish's time, the Islamic design templates had become more well entrenched (Mitra, 2002).

The Iron Pillar (Fig. 4.1; point 4) dating to the 4th century AD, standing in the inner courtyard of Quwwat-ul-Islam mosque, is another unique structure. It is conjectured that initially it was a standard (*Dhwaja Stambh*) outside a temple to support *garuda*, the vehicle of Vishnu. Some characteristic elements like the fluted "bell" capital with its *amalaka* form is reminiscent of Gupta architecture of the northern part of India (Page, 1926). A section of historians believe that it might have belonged to the Gupta Empire from where it was transported here by Anangpal, the Tomar king. It stands, austere, in the courtyard, having puzzled many a mineralogist (Cole, 1873; Spear, 1997).

4.2.3 Alai Darwaza and Alai Minar

The Alai Darwaza (Fig. 4.1; point 7) is a monumental gateway located south of the mosque. Built in 1311, it is most noteworthy of Alauddin

Figure 4.4 Red sandstone screen
(Photo: Gurmeet Kaur)

Khalji's extensions. The Alai Darwaza was the first example of a true dome built in India measuring 14 m high (Mitra, 2002). The single halled interior structure measures 10.5 m and exterior part measures 17.2 m and is 18 m tall and walls being a massive 3.4 m thick (Balasubramaniam, 2005). It was conceived as one of the four monumental gateways to the Quwwat-ul-Islam mosque, however, only this was completed. The *darwaza* with its arresting features like the pointed horseshoe arches (Fig. 4.5a) and decorative fringes of lotus buds add as an enhancement to the entry experience to the Quwwat-ul-Islam mosque. A *jaali* lined corridor runs on the one end of the outer door, throwing patterned beams of sunlight on the floor (Cole, 1873; Sharma, 2015). The openings are treated in marble lattice screens. Surface treatment in decorative floral tendril patterns is repeated with perfect symmetry on doorways. Intricate inscriptions in *Naksh* script adorn the domed gateway patiently crafted by Turkish artisans in red sandstone and marble inlays (Fig. 4.5b). Unique combinations of Saracenic and Hindu elements are fused to ornament the structure.

Alai Minar reflects Alauddin's desire to expand his power, and duly showcases the scale of his ambition. When he returned from his victories

Figure 4.5 (a) and (b) Intricate *Jaali* and relief work on the surface and interiors of Alai Darwaza

(Photos: Gurmeet Kaur)

Figure 4.6 Giant rubble masonry core, Alai Minar
(Photo: Gurmeet Kaur)

in Deccan, he desired to build another *minar* (Fig. 4.1; point 11) which was envisioned to be twice the length of the Qutb. However, it never progressed beyond the first stage and was summarily abandoned at his death in 1316 when only the first storey was built. Today it stands in a dilapidated condition, a giant rubble masonry core, like a stub, a prospect that could have outdone the Qutb (Sharma, 2015). Examining the dilapidated remains of the "ambitious" core structure, the rising radial trajectory of the windows clearly points to plans of having a gradual ramp as opposed to stairs in the case of Qutb, to ascend the minar from right to left. The partly built giant structure has a rubble masonry core using Delhi quartzite blocks which would have been intended to be clad in dressed stone once complete (Fig. 4.6; Page, 1926; Mitra, 2002).

4.2.4 Tombs of Iltutmish and Imam Zamin

The Tomb of Iltutmish is situated west of the extension he added to the mosque (Fig. 4.1; point 10). It was built by the man himself in 1235 and is strikingly different from the earlier part of the complex in that the advancement in architecture is markedly visible in the elements (the first squinch arch in India); the Hindu artisans had certainly developed a hold on constructing Persian elements independently, and also there are no more remnants of Hindu temples woven into the structures. The central chamber of the Tomb of Iltutmish has a square proportion of 9 m × 9 m. The outer as well as inner walls have elaborate carvings in stone with the main cenotaph being placed on a raised platform in the center of the square structure of the tomb. The interior sandstone walls are replete with intricate Quranic inscriptions in both *Naksh* and *Kufic* characters. However the dome built over the tomb could not survive, and now the structure lies under an open sky. Some scholars are of the view that it was originally conceived as a roofless structure, a practice connected with devout Muslims:

> Tombs are sometimes left uncovered, even when inside a mausoleum . . . the motive was a conviction that a grave not exposed to rain and dew was unblessed.
>
> (Dickie quoted by Spear, 1997)

The Tomb of Zamin was built in the 16th century (Fig. 4.1; point 8) and is clad with red sandstone, and the cenotaph is in white marble. The *mihrab* or the prayer niche wall facing Mecca is treated in white marble with detailed relief work carried out in stone. The culmination of Hindu motifs like the lotus, tassels, geometric diamond pattern and so on with Islamic sensibility is demonstrated beautifully (Mitra, 2002).

4.2.5 Alauddin Khalji's Tomb and Madrassa

Towards the rear side of the complex, an L-shaped block comprises of Alauddin's Tomb and a madrassa (Fig. 4.1; point 9). In dilapidated condition now, the central room where the tomb is situated has lost its dome. The remains of the tomb suggest that there was an open courtyard on the west and south side of the tomb with a north entrance. A series of cell-like rooms were meant to be classrooms of the madrassa. The outer walls of the madrassa incorporate corbelled pendentives that assist in transitioning from walls to the octagonal rim that is meant to support the dome (Page, 1926). The use of quartzite is common in Alauddin Khalji's Tomb and Madrassa (Fig. 4.7).

The Qutb Complex encloses myriad buildings dating from the 12th century consisting of the famous Qutb Minar, funerary and religious structures, *inter alia*, the Quwwat-ul-Islam mosque, Alai Darwaza and Alai Minar, capturing the march of history in volatile times. In totality,

Figure 4.7 Use of quartzite in the Alauddin Khalji's Tomb
(Photo: Gurmeet Kaur)

the Qutb Complex can be viewed as heralding the beginning of a new narrative in Indian aesthetics and architecture. From the destruction of the armies to the gradual evolution of a more distinct vocabulary, the Qutb Complex has been witness to it all. What started as a destructive encounter, sealed with signs of demolition and victory, gave way to a more peaceful play of distinct elements that, over time, fused in natural ways to produce a hitherto unseen aesthetic, the Indo-Islamic school of architecture which was to integrate disparate elements into a more developed distinct new language that was not only to reflect the remnants of the cultural influence of the new empire but also the raw materials, techniques and aesthetics of the existing lands.

4.3 Humayun's Tomb and adjoining monuments, Delhi

> So delicately carved, so fair,
> The graceful buildings stand,
> Such as to us are like the dreams
> Of some enchanted land.
> He looked upon them as the scrolls
> Prophetic of our life,
> The chronicles where Fate inscribes
> Our sorrow, sin, and strife;
>
> *The Tomb of Humaioon* (1832) by Letitia Elizabeth Landon

The nascent Islamic influences palpable in Quwwat-ul-Islam mosque and Qutb Minar reach a prodigious flowering in the architecture of Humayun's Tomb (Fig. 4.8). It also presages the splendor achieved in Shah Jahan's reign, particularly in the line of monumental mausoleums, reaching a high point with the Taj Mahal. Humayun's Tomb (Fig. 4.8; point 8), built in 1570, located in Nizammudin was accorded the status of UNESCO World Heritage Site in India in 1993. It was included under two criteria: (i) to represent a masterpiece of human creative genius and (iv) "to be an outstanding example of a type of building, architectural or technological ensemble or landscape which illustrates a significant stage(s) in human history" (https://whc.unesco.org/en/list/232).

A successor to Babur, the founder of the Mughal dynasty, Humayun's authority was challenged by the governor of Bengal, Sher Shah Suri, who set up the Sur dynasty in 1538. Humayun's love of wine and poetry and decadent ways were way too flamboyant for the task of empire consolidation. Within 10 years of his ascending the throne in 1530, Sher Shah overpowered him, forcing him to flee to Persia in 1540 for an interim. He returned in 1555 with a renewed might and an army of soldiers, poets, writers and builders. He was quick to reclaim control over his territories, but it turned out to be a rather ephemeral

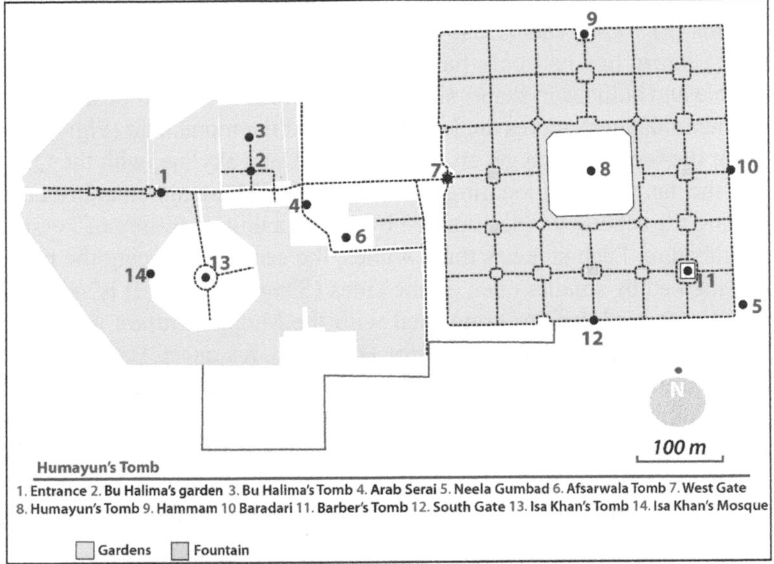

Figure 4.8 Plan of Humayun's Tomb Complex

(Source: Traced from Google map 2019)

glory. Immediately afterwards, he had a fatal fall at the stairs of his library and died instantly. His wife, Hamida Banu Begum or Begum Bara, in true epic grief, attended by royal excess announced in 1565 that she would not rest till she got a grand mausoleum erected in Humayun's memory on the banks of Yamuna River and that that was to be her ultimate tribute to her husband. Additionally, the site was close to the *chilla* of Nizam ud din Auliya, the revered Sufi master (Sharma, 2015).

4.3.1 Humayun's Tomb

The building is an example of a blend of Timurid Persian vocabulary and a pre-Mughal stamp of artistry. A bulbous double dome, meticulous plan and high portal front elevation – all characteristics of Timurid architecture – rub shoulders with intricate marble inlay work, lotus bud fringed arches, perforated stone *jaali* screen and wide *chhaja* eaves (Koch, 1991). It was one of the very first structures in India to use a double dome, a distinct feature that gave the building an imposing exterior while keeping the ceiling of the interior hall in proportion with the interior heights (Fig. 4.9). This is also the first full dome in India, which till the 15th century had had only half domes. The outer dome is covered in marble and bulbous in shape, supported by *chhatris* or pavilions, another element that underscores the hybrid nature of the monument (Fig. 4.10). From the roof, you can get an eyeful of the Delhi skyline, with the Qutb and the Jama Masjid looming on the horizon. The main chamber containing the cenotaph is surrounded by *jaalis*, a unique feature of Persian architecture. Each side has three arches, the central one being the highest, flanked by smaller ones on the sides (Sharma, 2015). It is said that about 150 dead persons connected with the Mughal Empire were buried in Humayun's Tomb – Emperor Humayun, his queen Banu Begum, Prince Dara Shikoh and high-ranking courtiers, though the disposition of their coffins is not known exactly. It is for this reason that the Humayun's Tomb is referred to as "Dormitory of the House of Timur" (Spear, 1997).

As compared to Babur's resting place in Afghanistan, it indeed is a much more grandiloquent structure. It is first in line of the garden tombs

Figure 4.9 Humayun's Tomb: the main mausoleum
(Photo: Parminder Kaur)

Figure 4.10 Double dome of Humayun's Tomb clad in marble and supported by *chhatris*

(Photo: Gurmeet Kaur)

that became synonymous with the Mughal dynasty in India. Begum Bara appointed Miyak Mirza Giyas, the Persian architect from Herath, who henceforth "provided India with its first dome in Persian tradition" (Gascoigne quoted by Spear, 1997). Humayun's Tomb stands somewhere in between: the tentative patching together of Persian influences with Hindu symbolism of the 12th century had, by now, given way to a consolidated architectural vocabulary where the Persian forms of dome, arcs and *iwans* set to achieve greater structural perfection in the later period. It is also the earliest example of a tomb located in the precincts of a *chaarbagh* – a quadrilateral garden based on geometrical precision and underlining a tangible imagining of the Persian concept of heaven. It is a fine amalgamation of Timurid influence and local traditions (Sharma, 2015; Koch, 1991). To have a tomb ensconced in the gardens symbolizing paradise, with symmetrical gardens presented an idyllic imagining of an afterlife. The garden occupies enclosed paved walkways "khiyabans" that terminate at subsidiary structures and gatehouses. The garden is enclosed within a 6 m high arcaded wall on three of the sides. Causeways 14 m wide run along the central axis with narrow water channels flowing along the centerline (Spear, 1997; Sharma, 2015; Nanda, 2017).

Humayun's Tomb is built with three kinds of stone – local Delhi quartzite, Vindhyan Sandstone and Makrana white marble (Naqvi, 2002). White plaster is also used in areas where it was meant to mimic white marble. Red sandstone lattice screens have white lime plaster infill to keep the spirit of materiality alive in all parts of the structure. What could have been a monotonous structure is made exquisite with the use of intricately interlaid white marble, coupled with pristine white domes (Nanda, 2017). The mausoleum structure has an octagonal plan inscribed in a square base with chamfered edges (Fig. 4.9). The built form is raised on a meter high podium which in turn rests on the 6.5 m high terraced wide platform. Each of the sides of this high terrace is pierced with 17 arches, and the edges are chamfered to add to the aesthetic of the otherwise square base (Naqvi, 2002; Nanda, 2017). The central arches on each face open on to an ascending staircase while the rest of the arches open into cells or tombs of other family members. The terrace itself is clad with red sandstone as flooring material.

4.3.2 Neela Gumbad

On the southeastern side stands an impressive tomb covered in blue tiles. Known as Neela Gumbad (Fig. 4.8; point 5), it is said to have been built in 1625 by Abdu'r-Rahim Khan Khan-i-Khanan and said to contain the remains of Fahim Khan, one of his faithful attendants (Sharma, 2015). The tomb structure is octagonal in plan (Koch, 1991) that rests on a 33.2 m × 33.2 m square base and is raised on a 1.5 m high plinth (Nanda, 2017). The structure is built in grey quartzite stone and plastered internally as well as externally. The exterior of the dome is clad with dark blue tiles, and a combination of blue and yellow tiles is seen on the drum of the dome. The dome is crowned by an inverted lotus with a red sandstone finial, and the inner part is supported on squiches with plastered surfaces and elaborately decorated Persian inscriptions with incised plastering (Nanda, 2017; Naqvi, 2002). According to Spear (1997), the original color of the tiles was a mosaic of green, yellow and blue, owing to which the dome was called *sabz burj* or the green dome. However, ASI replaced it with blue tiles in the 1980s in an attempt to revamp the precincts. The stone inlay technique that achieved perfection in Persia (where it was carried out in bricks) was achieved here in abundantly available stone.

4.3.3 South Gate and the West Gate

The enclosure of the tomb can be accessible via two gates. Each gate rests on a podium which is approached by a series of five steps. Even though the South Gate (Fig. 4.8; point 12) is closed for public access, the West Gate (Fig. 4.8; point 7) is now being used for public access. This is also accessed via five steps and is a double storeyed structure. It comprises a 7 m square central hall with square side rooms on the ground level and oblong shaped rooms in the first level. The façade has arched recesses measuring 15 m from floor level to parapet heights surmounted with small *chattris* 1.5 m square in plan (Nanda, 2017). Both the gates are made in red sandstone and quartzite (Fig. 4.11).

4.3.4 Bu Halima's Tomb and garden and Isa Khan's Tomb

Not much is known about the origin of Bu Halima's garden (Fig. 4.8; point 2), but the architectural elements like the rubble walls of locally

Figure 4.11 Use of red stone and quartzite in the tomb gateway (Photo: Gurmeet Kaur)

Figure 4.12 Bu Halima's garden: rubble walls of locally found quartzite stone

(Photo: Gurmeet Kaur)

found quartzite stone suggest that the construction belongs to the 16th century Mughal period (Fig. 4.12). A dilapidated structure on the north side is said to contain the grave of Bu Halima (Fig. 4.8; point 3; Spear, 1997; Sharma, 2015). To the south of Bu Halima's garden is situated Isa Khan's Tomb (Fig. 4.8; point 13) made in the memory of Isa Khan who was a noble man in the court of Sher Shah Suri. It is also designed in the archetypal garden tomb design with a central mausoleum on a plinth, surrounded by arches, set in an octagonal garden enclosure (Fig. 4.13).

4.3.5 Arab Serai, Hammam and Barber's Tomb

On the southwestern corner is the structure called Arab Serai, which it is believed was built by Bega Begum to accommodate the 300 mullahs (priests) she had invited from Persia (Fig. 4.8; point 4). The building is divided into two quadrangles, consisting of rows of cells. The north wall contains an arcaded pavilion on a raised plinth of 2.1 m. Towards the

Figure 4.13 Archetypal garden tomb setting of Isa Khan's Tomb

(Photo: Gurmeet Kaur)

center of the inner side is an octagonal tank about a meter across. The room appears to be a bath or Hammam (Fig. 4.8; point 9). It appears to be built more for functional purposes with no ornamentation. Towards the north of the enclosure wall lies a rubble built circular well that supplied water to both the Hammam as well as the channels of the *chaar-bagh* (Nanda, 2017; Naqvi, 2002). Barber's Tomb, also known as *Nai ka Gumbad* (Fig. 4.8; point 11), houses two unknown graves inscribed with Quran verses (Sharma, 2015). Clad in red and grey sandstone, it is perched on a podium 2.44 m high on a 24.3 m × 24.3 m square base reached by seven steps from the south side. The central tomb enclosure is surmounted by a double dome, which rises from a unique 16-sided rim and is further crowned by an inverted lotus finial. At the four corners are *chattris* that give hints of green, blue and yellow inlays (Nanda, 2017).

Humayun's Tomb is said to be the first of its kind garden tomb in the Indian subcontinent. It is more often than not compared to Saint Paul's Cathedral in London and Saint Peter's in Rome. It is also said that no earlier precedent of this nature was present in history anywhere in the world – in scale, grandeur, order and symmetry (Nanda, 2017). Humayun's Tomb, in some ways, marked the end of the austere style of Indo-Islamic architecture and beginning of a varied material palette as seen in the Taj and so on.

4.4 Red Fort Complex, Delhi

Construction of Red Fort began in 1639 was initiated as a palace fort of Shahjahanabad, the capital and power center of the Mughal Empire under the fifth monarch, Shah Jahan. Adjacent to the fort of Salimgarh, built by Islam Shah Suri in 1546, multiple structures constitute the Red Fort complex (Fig. 4.14). The Red Fort was inscribed in the UNESCO list in 2007 under the following criteria: (i) to exhibit an important interchange of human values, over a span of time or within a cultural area of the world, on developments in architecture or technology, monumental arts, town-planning or landscape design; (ii) to bear a unique or at least exceptional testimony to a cultural tradition or to a civilization which is living or which has disappeared; and (iii) to be directly or tangibly associated with events or living traditions, with ideas, or with beliefs, with artistic and literary works of outstanding universal significance (https://whc.unesco.org/en/list/231).

The name "Red Fort" alludes to the use of red stone in erecting its imposing outer walls and structures (Fig. 4.15). The fort, which constituted the living quarters of the ruling empire for 200 years, represented a high point of refinement in art and architecture spearheaded by Shah Jahan. It is for this reason that the fort, constructed next to the Yamuna, acquired the moniker "Qila-i-Mubaraq" or the Blessed Fort. Together with the adjacent fort of Salimgarh, built by Islam Shah Suri in 1546, the precinct is referred to as Red Fort Complex. The construction started in 1639 and culminated in 1648, though additions kept happening as late as the 19th century.

The Red Fort was designed by Ustad Hamid and Ustad Ahmed (Spear, 1997; Sharma, 2015).

The Red Fort has a similar context as the Agra Fort, as it lies along the river Yamuna which also originally replenished the moats around the fort. The course of Yamuna has now changed, and therefore the river is substituted with a road. The eastern side faced the river while the other three sides were protected with a 9.1 m deep moat (Hearn, 1906; Sanderson and Shuaib, 2000). The fort is octagonal in plan with its north-west axis being longer than the east-west axis. The total area covered being 1 sq km enclosed by approximately 2.41 km of fortification peripheral walls

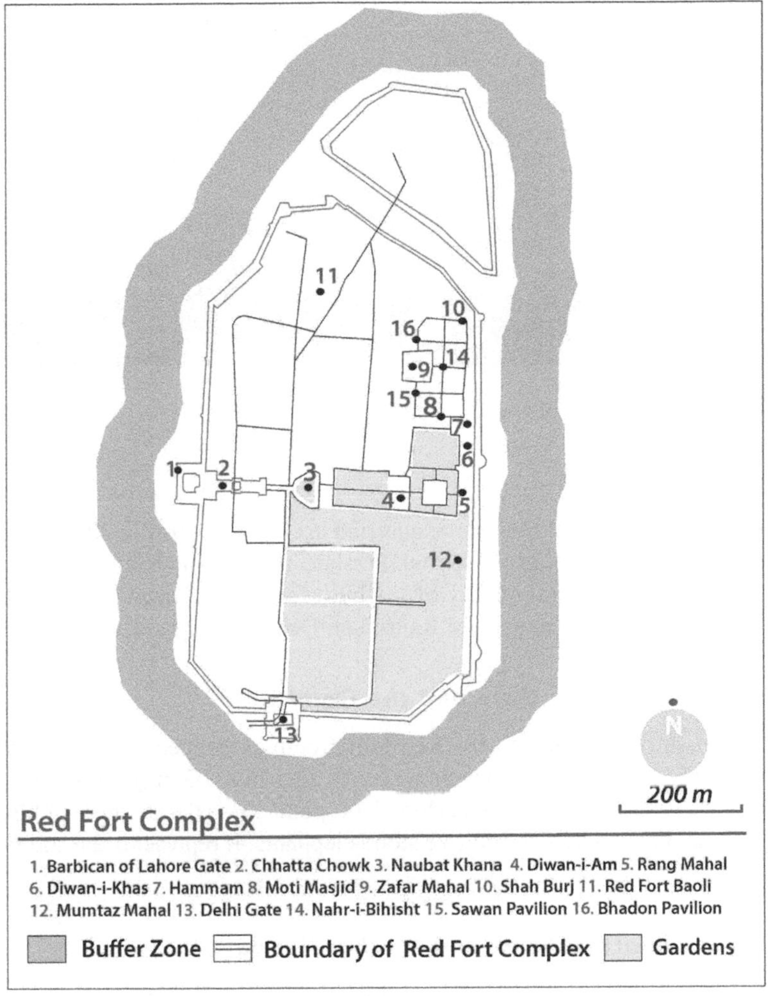

Figure 4.14 Plan of Red Fort Complex

(Source: Traced from Google platform 2019 and https://whc.unesco.org/en/list/231)

punctuated with turrets and bastions of varying heights of 18 m on the river front and 33 m on land front. The primary building material used is red sandstone along with limited use of Makrana white marble. Surface ornamentation was done with *pietra dura*, mirror work and the use of

Figure 4.15 Red Fort
(Photo: Gurmeet Kaur)

white plaster derived from stone quarried in Gujarat (Rezzavi, 2010). The style is a amalgamation of Mughal, Persian, Timurid and Hindu architecture. The architectural quality of the built form had a strong impact on what was built later in parts of Rajasthan, Delhi and Agra (Spear, 1997).

4.4.1 The Lahore Gate and the Delhi Gate

The main public entry to the Red Fort Complex happens from Lahore Gate (Fig. 4.14; point 1) which measures 12.5 m × 7.3 m (Hearn, 1906; Verma, 1985). The Delhi Gate (Fig. 4.14; point 13) is the southern public gateway flanked by two life size stone elephants. It represents the southern public entry to the fort.

4.4.2 Chhatta Chowk and Naubat Khana

From the Lahore Gate side, one could enter a *souq*/vaulted arcade called Chhatta Chowk (Fig. 4.14; point 2) with bay shops lining the passage. It was originally called Meena Bazaar or *Bazzar-e-Musakkaf*. This is a "Covered Bazaar" and was inspired by a similar structure in Peshawar. It is said that the shops stocked some of the finest jewels, textiles, silks, brocades and silverware and other exotic wares like enamel art, paintings and carpets (Hearn 1906; Verma, 1985). In the middle of the arcade 70 m × 8.2 m wide, is an octagonal court that has a sky-cutout where some remains of the original decoration of incised plaster are visible. Each side of arcade sports 32 arched units that were originally shops (Spear, 1997; Rezzavi, 2010; Sharma, 2015). The Naubat Khana (Fig. 4.14; point 3), also known as the Nakkar Khana/Drum House, provided access to the courtyard (Koch, 1991). The courtyard was the hub of musical activity and performance. During the high point of the Mughal Raj, the imperial band is said to have performed here six times a day (Sharma, 2015).

4.3.3 Diwan-i-Am

On the far side of the Naubat Khana is the Diwan-i-Am, (Fig. 4.14; point 4) the Public Audience Hall. It was a space primarily used for state functions and public meetings and addresses. The arcaded colonnades form the structural components of the hall made of red sandstone which were originally rendered in white shell lime plaster (Fig. 4.16). The hall is divided into three aisles of 7 feet each (Verma, 1985). The ceilings and columns were gilded, and railings of gold and silver created a hierarchical distinction of the visitors attending the sessions here.

Figure 4.16 Diwan-i-Am

(Photo: Gurmeet Kaur)

According to Spear:

> The hedges mark the positions of the old walls. The Diwan-e-Aam was covered with white plaster or chunam. Within the pavilion, the nobles stood in rows facing each other, according to their rank. The royal princes stood next to the throne and the *wazir* sat on the marble *takht* below it . . . the lesser nobles stood outside the Hall. Behind the emperor's throne there was some mosaic work done by a French artist. . . . One of the pictures is of a man playing a violin. This represents Orpheus, the Greek God of Music. These stones were taken away during the mutiny of 1857 but Lord Curzon discovered them in London and put them back here. In the hot weather great red curtains were hung around the hall to keep off the sun . . . the reason why Mughals built their halls only with pillars and no walls was that they came from Central Asia where they always lived in tents. When the Mughals came to India they built in stone but still thought of their tents. The Diwan-e-Aam is *shamiana* in stone.
>
> (Spear, 1997)

4.4.4 Mumtaz Mahal, Rang Mahal, Khas Mahal, Nahr-i-Bahisht and Diwan-i-Khas

To the south of the main structure were the *zenana* quarters, Mumtaz Mahal (Fig. 4.14; point 12) and Rang Mahal (Fig. 4.14; point 5). The latter is so named for its use of vibrant color decoration and mosaics in the interiors. The "Palace of Color" or Rang Mahal was meant to house the emperor's many wives. It was embellished with colorful mosaics of mirrors, gilt and color. The central hall was divided into 15 bays with ornamental pillars. The central pavilion is fed by Nahr-i-Bihisht (Fig. 4.17; Spear, 1997). The private apartments are connected by a continuous water channel, known as Nahr-i-Bihisht (Fig. 4.14; point 14) or the "Stream of Paradise," a concept generally seen in the garden setting of Islamic architecture to emulate "paradise" as described in the Quran. This stream of water runs from the north passing through built forms and flowing towards the south. Water was drawn from the river Yamuna via a tower known as the Shah Burj (Fig. 4.14; point 10). Its pavilions lie in the eastern edge of the fort. A doorway northward of the Diwan-i-Am leads to the *Jalau Khana* (Diwan-i-Khas) which is the innermost court of the palace (Fig. 4.14; point 6). The pavilion is supported on white marble pillars, and engrailed multi-foliated arches divide the hall in 15

Figure 4.17 The central pavilion is fed by Nahr-i-Bihisht
(Photo: Gurmeet Kaur)

bays (Hearn,1906; Verma, 1985). Floral inlays on the piers and perforated tracery work add to a rich spatial experience (Fig. 4.18a). It is said to have sported a silver ceiling (which now is restored in wood). The center of the east wall adorned a grand peacock marble throne which was exquisitely detailed in a curving Bangalda or whaleback roof and floral carvings on the lower front part of the throne structure. Today the lone platform upon which it once stood exists. The part of eastern wall forming the backdrop of the throne is decorated with *pietra dura* showing flora and fauna (Fig. 4.18b). The space between Diwan-i-Am and Rang Mahal was made into a garden (Spear, 1997).

4.4.5 Sawan, Bhadon Pavilion, Hayat Baksh Bagh and Zafar Mahal

The Hayat Baksh Bagh is a garden in the northeast direction of the main fort. There are two pavilions called Sawan and Bhadon (Fig. 4.14; point 15 and 16), named after the twin Indian months of rains and monsoons.

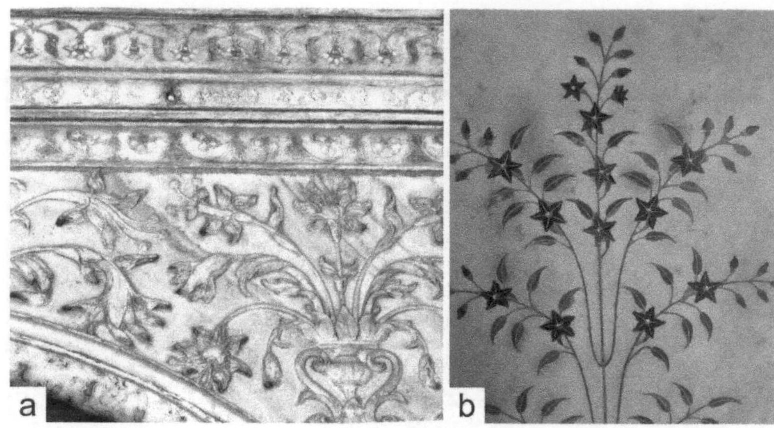

Figure 4.18 (a) Floral inlay work in Diwan-i-Khas and (b) *Pietra dura* showing flora and fauna in Diwan-i-Khas

(Photos: Gurmeet Kaur)

Figure 4.19 Zafar Mahal

(Photo: Gurmeet Kaur)

The Zafar Mahal (Fig. 4.14; point 9) in red stone was added here by Bahadur Shah Zafar around 1842 (Fig. 4.19). The Mehtaab Bagh and smaller gardens existed to the west but were destroyed by the British (Sharma, 2015).

4.4.6 Moti Masjid, Hammam and Baoli

The Moti Masjid (Pearl Mosque) was a subsequent addition by Aurangzeb, primarily for his personal use, in 1659 (Fig. 4.14; point 8). It is a three domed marble structure with an arched screen (Fig. 4.20). The three arched screen leads to the courtyard, and the mosque is enclosed in a high peripheral wall that masks most of the mosque from outside leaving the view of the domes. The domes were originally covered in gilded copper plates but got damaged in 1857 and now exist as pure white. The floor is inlaid with prayer rug outlines (*musallas*) in black marble (Sharma, 2015). The Hammam (Fig. 4.14; point 7) was the royal bath complete with marbled floors (Sharma, 2015). The Baoli (Fig. 4.14; point 11) is believed to be a pre-Red Fort construction. It was converted into prisons by the British Administration. The stepwell or *baoli* is a beautiful piece

Figure 4.20 Moti Masjid
(Photo: Anuvider Ahuja)

of architecture built in a monolithic expression of stone, even though at the time of its construction, the aesthetic must have been driven by its functional requirements of providing for two sets of staircases to reach the well. The sets of arches on all four sides with the geometric quality of steps are fascinating albeit with a play of solids and voids.

The Red Fort reflects a fusion of Hindu, Timurid and Persian influences. The spatial arrangements and elements like the Mughal gardens were to have a long lasting impact on the subsequent architecture developed in the region. Nadir Shah stripped the Red Fort of its inlay work and jewels during his invasion of Delhi in 1739. Similarly the British destroyed its marble structures during the tempestuous revolt of 1857. While the Red Fort symbolized the zenith of the Mughal Empire, it also ironically became a site of its decimation in that it was here that the last emperor, Bahadur Shah Zafar, was tried before being pensioned off to Yangon in 1858. The fort was then turned into a garrison by the British. It is also a multi-layered structure, in that every sub-stratum encloses a layer of influence. The Red Fort singularly remained and continues to be a symbol of the myriad occupations, dynasts and power centers that claimed and reclaimed Delhi as their nucleus of power. The British added new fortifications to the original structure and till recently the premises enclosed an Indian Army Unit.

It is from the parapets of Red Fort that the declaration of Indian Independence was made and continues to be reaffirmed with the Prime Minister's speech delivered from its imposing ramparts every year on 15 August. The tradition goes back to the unfurling of the tricolor from the Lahore Gate on the day of Independence in 1947.

4.5 Taj Mahal and adjoining monuments, Agra

> You knew, Shah Jahan, life and youth, wealth and glory, they all drift away in the current of time. You strove, therefore, to perpetuate only the sorrow of your heart. . . . Let the splendour of diamond, pearl, and ruby vanish like the magic shimmer of the rainbow. Only let this one tear-drop, this Taj Mahal glisten spotlessly bright on the cheek of time, forever and ever.
>
> O King, you are no more. Your empire has vanished like a dream, your throne lies shattered . . . your minstrels sing no more, your musicians no longer mingle their strains with the murmuring Jamuna. . . . Despite all this, the courier of your love, untarnished by time, unwearied, unmoved by the rise and fall of empires, unconcerned with the ebb and flow of life and death, carries the ageless message of your love from age to age: "Never shall I forget you, beloved, never."
>
> Rabindranath Tagore (translated by Roy, 1966)

Inscribed in the list of World Heritage by UNESCO in 1983, the Taj Mahal at Agra (Fig. 4.21) is therein described as "the jewel of Muslim art in India and one of the universally admired masterpieces of world heritage." It is enlisted under criteria (i) which "represents a masterpiece of human creative genius" (https://whc.unesco.org/en/list/252).

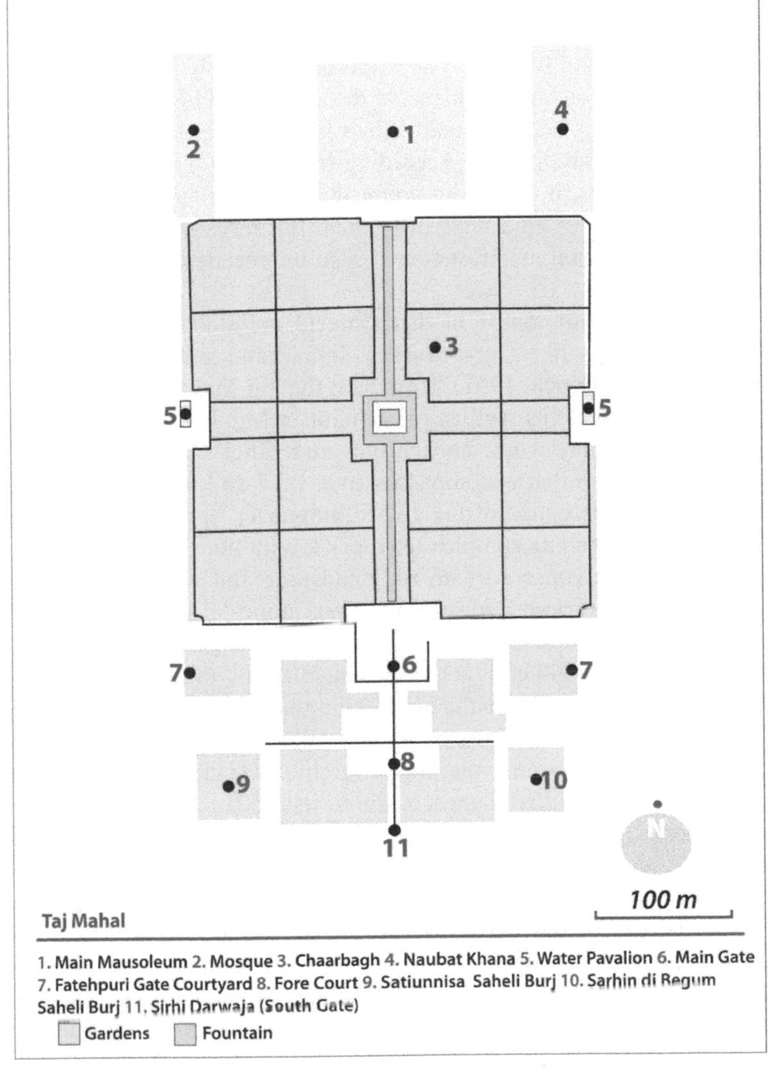

Figure 4.21 Plan of Taj Mahal Complex

(Source: Traced from google platform 2019)

4.5.1 Taj Mahal

Taj Mahal has had such a hold on popular imagination that it symbolizes love for most and the idea of India itself to many. Its soaring domes and minarets reflected in its four rivers of paradise represent the height of architectural beauty during the reign of the Mughals (DuTemple, 2003). Built with white Makrana marble, the monumental tomb was commissioned by Emperor Shah Jahan in 1632 in the memory of his beloved wife, Mumtaz Mahal, who had according to legend and lore, died birthing the emperor's 14th child. The emperor was deeply grieved and commissioned the monument to preserve the memory of his wife and accord her a fine resting place. Eventually Shah Jahan's Tomb too was enshrined here after his death in 1666. According to Nath (1972): "Among sepulchral monuments, it is without doubt the most brilliant and successful achievement." This epic scale of grief in the service of romantic love and its monumental manifestation is also unprecedented in the history of ideas.

The building of tombs in the Timurid architecture has progenitors in buildings like Gur-e-Amirin Samarkand and the Humayun's Tomb in Delhi (Koch, 1991). However, the Taj Mahal symbolizes the high point of scale as well as refinement in Mughal architecture and allied arts of inlay work, horticulture and relief work. Set amidst a grand *chaarbagh* that encompasses over 0.17 sq km, The Taj Mahal is located on the banks of the river Yamuna in Agra, Uttar Pradesh. The Taj Mahal and its complex used brick with lime mortar as its core building blocks veneered with red sandstone and white marble with inlay work of precious and semi-precious stones. The construction of the main structure started in 1632 and was completed in 1648. The two structures on either side of the mausoleum – the Mosque and the Guest House – as well as the south side main gateway and the outer court were phased subsequently and executed in 1653 by a work force of 20,000 laborers under the master architect Ustad Ahmad Lahauri (Nath, 1972) (https://whc.unesco.org/en/list/252).

The tomb (Fig. 4.21; point 1) is placed at one end of the quadripartite garden instead of the center, adding visual depth when witnessed from a distance (Nath, 1972; Siddiqi, 2009). The one stunning feature of the Taj Mahal that stands out is its inviolable symmetry (Fig. 4.22). Not only is the main mausoleum a perfect structure in terms of manifesting a symmetry that is well nigh perfect but also the buildings flanking the Taj have been so placed as to enhance this aspect as the fundamental

Figure 4.22 The main mausoleum in Makrana marble, topped with a magnificent large dome and four *jharokhas*

(Photo: Anuvinder Ahuja)

principle of design. The Taj Mahal, too, displays this principle to a tee, by being identical as a vision, as a vista, no matter from where the view is framed.

The architectural quality of the Taj Mahal operates at two levels: one is of the grand monolithic structure raised on a square plinth of a marvellous block of accurate masonry 95.4 m × 95.4 m and 5.61 m high, which is accessed via a flight of lateral steps on the southern side (Nath, 1972; Siddiqi, 2009). The structure, clad in Makrana white marble, is square in plan with its chamfered corners (Fig. 4.22; Nath, 1972; Siddiqi, 2009). The building is approachable through massive arched gateways called *iwan*. The ivory marble gateways are decorated with the finest stucco, geometrical designs and inlay with gem stones (Fig. 4.23). Bands of Arabic calligraphy in showcasing *ayats* from the Quran frame the four *iwans* (Fig. 4.23). On each side are four small *iwans* on the ground level and

Figure 4.23 Inlay work and bands of Arabic calligraphy showcasing *ayats* from the Holy Quran frame the *Iwan*

(Photo: Anuvinder Ahuja)

upper level. Four minarets stand sentinel on sides of the Taj, visible from all vantage points, without obstruction.

The second engagement with the Taj is at an intimate level which is an equal measure of tangible and that of intangible elements. As one enters the central chamber which houses the cenotaphs of Mumtaz Mahal and Shah Jahan, the surface treatment seen therein is done using the most exotic and decorative aesthetic techniques. The interior walls are laced with delicate filigree *jaali* work on the outer walls, ensuring a stream of filtered light falling on the ornamental (though false) sarcophagi of Mumtaz Mahal and Shah Jahan. These are enclosed with *jaali* walls exhibiting the finest precious and semi-precious stones in *pietra dura* and lapidary work. The exquisite materiality is laced with the poetics of the space creating a complete magical experience which is beyond tangible parameters of architecture and design. Even though the cenotaphs are of symbolic relevance, the real graves lie in the lower chambers of the crypt, a practice that was commonly deployed in that Mughal era.

The building is topped with a magnificent large dome and four *jharokhas* with identical smaller domes (Fig. 4.22). The upper part of the dome is decorated with lotus design that accentuates the shapely ascent. The domes and *chhatris* are topped with gilded finials that showcase an amalgamation of Persian and Hindu elements. A crescent, typically Islamic motif rests atop the central finial. The minarets at the corner of the platform add a spatial reading and a third dimension of the monument that has become synonymous with the imagery of the Taj. Four minarets are designed for muezzin calls, thus integrating the functionality of a mosque into the mausoleum. These three tiered circular minarets measure 5.6 m at the base and taper upward to 3.54 m in diameter. The entry to each minaret is via a door 1.83 m high and 0.79 m wide also opening out on the first and second floor to their respective balconies. The minaret throughout its height has a solid core structure that brings the structural stability to each minaret. Around the solid core winds a staircase from the base of the balcony to the third storey comprising of 51 steps in the first storey, 40 in the second and 58 steps in the third storey with a riser of 0.2 m (Nath, 1972; Siddiqi, 2009). The four sides of the octagonal base of minarets extend beyond the edges while the central doorways, arches, concave and convex surfaces, light and shadow and a play of solids and voids all blend beautifully in a magnificent Mughal garden setting (https://whc.unesco.org/en/list/252).

4.5.2 Mosque and Naubat Khana/Guest House

The Taj Mahal Complex houses two almost identical buildings i.e. the Mosque (Fig. 4.21; point 2) and the Guest House (Fig. 4.21; point 4) which mirror each other on either side of the tomb. On the west side of the Taj is the Mosque in red sandstone. Its main design is that of an expansive hall surmounted by three domes (Fig. 4.24). The exquisite gateway and gleaming floor of the Mosque lined with prayer mat motifs in red sandstone frame a magnificent view of the Taj. An identical building, a Naubat Khana (Guest House), flanks the Taj Mahal on its eastern side and with some penchant is called “jawab” – as in an “answer” to complete the symmetry set forth by the Mosque. Both buildings have identical facades but functionally differ; for instance the Mosque has *mihrab* – a prayer niche in the wall facing Mecca. Both structures, being red sandstone, have an elaborate ornamentation with white marble inlays (Fig. 4.24).

Figure 4.24 Marble inlay in Red Stone characterizes the workmanship of the Mosque

(Photo: Anuvinder Ahuja)

4.5.3 Main gateway

This imposing double storeyed gateway (Fig. 4.21; point 6) lies almost in the center of the southern perimeter boundary (Fig. 4.25). Standing on a near square plan and over 100 feet in height, it is flanked by octagonal shaped towers topped with elegant cupolas imbibed in a *chattri* design. The façade treatment is typically Islamic in its approach and perfect mirrored symmetry with recessed arches with calligraphic bands of marble inlays (Fig. 4.26). An exquisite *thulth* (triangular style of calligraphy) script adorns the white bands with one of the verses that says *Wal Fajr* meaning "now enter the paradise." The divisions of the garden in front of the galleries by sub-dividing into four quartets is characteristic of the Timurid and Persian principles of sub-divisions (Siddiqi, 2009).

4.5.4 Chaarbagh

From the entrance, the tomb is approachable through a green expanse of Mughal gardens which are modeled after the Persian design. As

Figure 4.25 The imposing double storey south gateway/*Iwan*
(Photo: Anuvinder Ahuja)

discussed earlier, in the Islamic worldview, the *chaarbagh* were four rectangular gardens set amidst four flowing rivers of *Jannah* (paradise). During the Mughal reign, the gardens boasted a profusion of local flowers and vegetation which included a rich horticultural spread of roses, ornamental and fruits trees (Koch, 1991). The Taj Mahal Complex is set in a grand 300 m square *chaarbagh* (Fig. 4.21; point 3). Most Mughal gardens are rectangular layouts with the pavilion placed in the center but in the case of Taj Mahal, they deviated from the norm and placed the tomb at the far end of the garden. One of the interpretations that is most fascinating is that the creators of the Taj evolved the idea of the *chaarbagh* taking the immediate context of the river Yamuna into the garden design and assuming it to be the "river of paradise." The avenues of cypress trees line the path on either side of the water channel culminating at the main gate. The garden assumed significant development under Shah Jahan (Fig. 4.27; Koch, 1991; Nath, 1972; Siddiqi, 2009). On the east and the west sides of the gardens are located two water pavilions at the two ends of the broad water canal (Fig. 4.21; point 5).

Figure 4.26 Calligraphic bands on Iwan
(Photo: Sakoon Singh)

During the rebellion of 1857, Taj Mahal was vandalized by the British soldiers and precious stones taken away. Lord Curzon ordered a restoration of the Taj and that included the installation of an ornate hanging lamp and sprucing of the gardens that had fallen into disuse. However, when the British took over they attempted to, *inter alia*, model the

Figure 4.27 The avenues of cypress trees line the path on either side of the water channel culminating at the main South Gate

(Photo: Noor Dasmesh Singh)

gardens according to the English aesthetics, making them more uniform and devoid of ornamentation characteristic of Mughal times (Herbert, 2012). In fact, the lore suggests a purported attempt to even demolish the Taj (Spear, 1949). However, thc Taj survives today as a monument encompassing the original intent of its conception.

Shah Jahan's genius in discerning art and vision drawn up with his architect Ustad Ahmad Lahauri and Mir Abdul Karim created a grand, timeless monument. Mounted on a high plinth, the Taj throws a reflection in the waters that seems as majestic as the monument itself. The effect of looming shadows of the night, the profusion of milky whiteness of full moon light as it scatters on blocks and expanse of soft marble, coupled with the clear stream create an imagery that is ethereal.

4.6 Agra Fort and its monuments, Agra

> "*Marble, I perceive, covers a multitude of sins.*"
>
> – Aldous Huxley

Agra Fort, at Agra, spread over 0.38 sq km, has had a long, checkered history, and it would be more apt to describe it as a mini township than a lone monument (Fig. 4.28). There are layers of deep historical time, imprints of which are spread throughout the complex. The Agra Fort was inscripted by UNESCO in 1983 under criteria (iii) to bear a unique

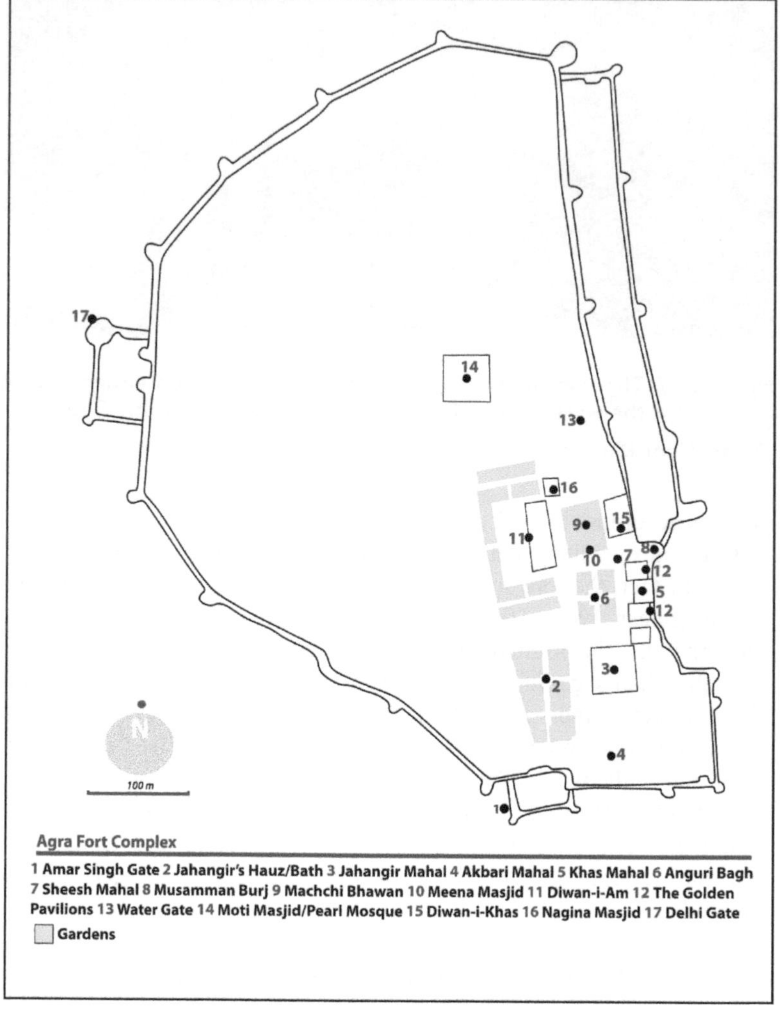

Figure 4.28 Plan of Agra Fort Complex

(Source: Traced from Google platform 2019)

or at least exceptional testimony to a cultural tradition or to a civilization which is living or which has disappeared (https://whc.unesco.org/en/list/251).

Agra Fort, also referred to as the "Red Fort" of Agra, is adjacent to the older fort called the Salimgarh, together with which it forms the Agra Fort Complex (Fig. 4.28). This mighty fortress encompasses a 2.5 km long and about 21.3 m high peripheral fortification. Agra Fort, in its rudimentary form, first as a post and subsequently as a symbol of victory, documenting a litany of possessions and dispossessions through history, has stood vanguard since its ancient name as Badalgadh. Mahmud of Gazhnavi conquered the fort in the 11th century. The Lodhi dynast Sikandar Lodi (1489–1517 AD) moved his capital from Delhi to Agra and occupied it in addition to constructing other landmark buildings. After the First Battle of Panipat (1526), the Mughals captured it and it is here that the coronation of Humayun took place. In 1558, Akbar completely remodeled the existing structure which was made of red brick. Shah Jahan, with his love of marble, remodeled the fort and got some buildings erased and new ones erected. However, it was here, ironically, that he was later imprisoned by his own son, Aurangzeb. He died a sad man, occupying the Musamman Burj with its marbled parapet and affording him a view of the Taj, and a memory of his departed wife. The fort came under the imprint of the Jat rulers who built Ratan Singh ki Haveli. The Marathas held the fort during their ascendance in the 18th century till they were routed by Ahmad Shah Abdali in the Third Battle of Panipat. The Marathas regained control of the fort under Mahadji Shinde in 1785 till the British occupied it after the Second Anglo Maratha War in 1803. It became a site of battle during the Mutiny of 1857, and the fort walls still bear the imprints of it being vandalized in this turmoil of being passed from one occupant to another. In a way, the history of Agra Fort well nigh mirrors the history of Agra city itself (Siddiqi, 2008).

As per Abul Fazl's account, Agra Fort has some 500 buildings built in Bengal and Gujrat style. Not many buildings of the era exist anymore, and the count is now given as 27 (Siddiqi, 2008). Rendered in red sandstone, this imposing fort houses many smaller palaces, two mosques and other structures inter-woven in a beautiful cohesive urban plan (Siddiqi, 2008; Verma, 1985). The fort is conceived as a sketchy semi-circular plan with its straight edge lying parallel to the course of Yamuna river. The core structure in most cases is brick with lime mortar which was veneered with either red sandstone or white marble. The decorative elements used various forms of inlays with precious and semi-precious stones. Finishes also include stuccoed white to imitate white marble finish (Havell, 1904; Siddiqi, 2008). Owing to a strategic use of its architectural features, Agra Fort was supposed to be impregnable. Its double ramparts are strengthened with circular bastions at regular intervals and a deep moat encircling the entire length of the boundary wall (Fig. 4.29). It is with a guarded irreverence and in an attempt to trivialize the enemy that the British official describes the structure of the Agra Fort, which he dismisses as a structure with "no real strength," and then in a nervous build up, goes on to describe the details:

> Opposite to the Jama Masjid, across an open square, stands the Fort, whose walls are 70 feet high and a mile and a half in circuit; but as they are only faced with stone and consist within of sand and rubble, they have no real strength, and would crumble at once before the fire of modern artillery. A draw bridge leads across the deep moat which surrounds the crenellated ramparts, giving access through a massive gateway and up a paved ascent to the inner portal. The actual

Figure 4.29 Moat encircling the entire length of the Agra Fort, with Taj Mahal looming in the view

(Photo: Sakoon Singh)

entrance is flanked by two octagonal towers of red sandstone, inlaid with ornamental designs in white marble.

(Frowde, 1908)

4.6.1 Amar Singh, Delhi and Water Gate

This red sandstone gate to the south of the fort was initially called Akbari Darwaza, which is also the public visitor entry (Fig. 4.28; point 1). It was renamed Amar Singh Gate by Shah Jahan after his trusted Rajput Lieutenant (Fig. 4.30). The main entrance leads to a colonnaded court area. The bastions are covered by *chhatris*, circular *chhaja* and inverted lotus copulas. These were decorated in Persian style with glazed tiles that have all but withered, except for a few places (Verma, 1985; Siddiqi, 2008). A sharp diversion for protection purposes is created leading up to a "ramp entrance" paved with red bricks. This further leads to a red sandstone gateway structure on the eastern side, which is the next level of entry hierarchy. A central archway acts as an opening and is decorated with two-part arched panels with marble inlay ornamentation. The top

Figure 4.30 Amar Singh Gate, now used as the public entry for the Agra Fort (Photo: Anuvinder Ahuja)

part of the gate has six rectangular loopholes over which three loopholes bear temple *shikar* form made out of red sandstone. The topmost part has seven parapet merlons that gracefully terminate the structure (Siddiqi, 2008). To the west of the fort is located the grand Delhi Gate (Fig. 4.28; point 17). Built by Akbar in 1568, it was meant to be the king's personal entrance and grandest of all gateways. Embellished with fine inlay work, it became a combination of strength, yet grace. Access to the fort through Delhi Gate is closed these days because it falls in the part under the Army area. The Water Gate (Fig. 4.28; point 13) with its front towards the river front lined with *ghats* (quays) is called Khizri Gate or the Water Gate. Given the vulnerability of this side to imminent attacks, it was a formidable structure.

4.6.2 Akbari Mahal

What now appears like a brick-clad, almost open courtyard was at one point the Akbari Mahal (Fig. 4.28; point 4). The plan of the Mahal can still be traced by following the remains of the structure that is located northward of the Jahangir Mahal. This being the oldest building of the Agra Fort complex together with the stepped well *baolis*, the construction appears to be post and lintel type with excessive use of red sandstone to create a series of rooms with double storied pavilions as an extension of the rooms on the eastern side. Stepped lavatories built in solid stone slabs for the ladies of the harem also exist as part of the structure (Hussain, 1937; Siddiqi, 2008).

4.6.3 Jahangir Mahal and Jahangir's Hauz/Bath

Contrary to its name, Jahangir Mahal was commissioned by Akbar rather than Jahangir (Fig. 4.28; point 3). Placed in the southeast part of the master plan, this interesting piece of architecture is rather austere on the outside but is heavily detailed on the inside with relief work and exquisite surface ornamentation. The core structure is *Lakhauri* bricks clad over by finely dressed red sandstone. The almost flat symmetry of the façade is punctured by projecting porticos supported on circular columns and elegantly carved brackets, the top part of which is supported by slanting planes of linear shading elements. Use of white marble is made to create a sort of a pattern on the façade as a means of ornamentation (Siddiqi, 2008). The entry into the central *Deorhi* (portico) leaves one awe-struck with the beautiful relief work, carved out lotus petals, parrots and other fauna appointed on the four arches with lotus inspired frieze and squinch details. The inner colonnade with deep stone clad columns supports the exquisitely detailed elements that join together to form unique pattern of arches (Fig. 4.31). These are structurally supported by finely crafted stone brackets that line the inner surface of the double storey structure after regular intervals. The covered part of the colonnade is raised on a plinth to give it a certain hierarchy (Verma, 1985). The Jahangir Mahal presents some admirable examples of Hindu carving, with projecting brackets as supports to the broad eaves and to the architraves between the pillars, which take the place of arches (Fig. 4.31; Koch, 1991). This Hindu form is adopted in the Jahangir Mahal and in the neighboring Saman Burj instead of the arch, and the ornamentation of the former is purely Hindu. It is here that he installed what was called the Chain of Justice (Zanjir-i-Adl) in 1605. It was a quintal heavy chain fashioned from 60 pure gold balls, a proof of the justice loving stance of Jahangir. It was supposed to be accessible to people at large, and anyone with a grievance had to just pull it. One end was tied to Shah Burj and the other end at river Yamuna for people to use (Siddiqi, 2008). Jahangir's Hauz or bath is a saucer shaped, stone bath. It is 1.22 m in height and 2.4 m in diameter (Fig. 4.28; point 2). On the outer side is an inscription that says "Hauz-i-Jahangiri" or bath of Jahangir. It has a dainty flight of steps leading to water. This elliptical piece is similar to a modern day

Figure 4.31 The inner court of Jahangir Mahal with carved projecting brackets

(Photo: Parminder Kaur)

"free standing bathtub" not meant to be embedded in ground, but to be placed on ground perhaps more like a piece of furniture. The bath has a seal carved as relief mentioning the year 1611 which coincides with the marriage of Jahangir and Nur Jahan (Siddiqi, 2008).

4.6.4 The Khas Mahal, Anguri Bagh, Musamman Burj, Sheesh Mahal and Diwan-i-Khas

The Khas Mahal (Special Palace) or Aaramgrah-e-Mualla was built between 1631 and 1640 by Shah Jahan (Fig. 4.28; point 5). Made of pure white marble, it was decorated with studded floral motifs which have now faded. It overlooks the Yamuna on the eastern side and on its west is the Anguri Bagh (Fig. 4.28; point 6), a garden in the quadrilateral plan with water channels, fountains and cascades. The palace is overlooked by red sandstone pavilions, covered in white plaster, which feature curved cornices derived from vernacular Indian prototypes. In all likelihood, grapes were grown here, giving it the moniker of Anguri Bagh. The large rectangular garden added by Shah Jahan in 1637 in front of Khas Bagh is patterned on the *chaarbagh* design. The hexagonal paths of the garden are treated in red sandstone interspersed with the grape vines creating poetics around architectural elements. The four gardens, measuring 5.5 m in width each are punctuated by white marble paths conjoined in a marble tank that sits in the center. The garden is flanked by a double storied structure of red stone painted with white plaster. Most likely these rooms constituted of the living quarters of the harem inmates (Hussain, 1937; Siddiqi, 2008). The entire court measures approximately 67.6 m × 52 m, and right below the east facing marble terrace lies a platform measuring about 14.7 m square (Hussain, 1937; Verma, 1985; Koch, 1991).

Musamman Burj sits atop the eastern battlement of the Agra Fort that faces the river and also has a stunning view of the Taj (Fig. 4.28; point 8). Shah Jahan, with his love for marble, is said to have had this rebuilt in marble between 1632 and 1640. It afforded him a view of the sun and the Taj and also an audience with the people from the lookout of the *jharokha* as was a common practice, attested through various depictions in painting. This complex connects with Diwan-i-Khas, Sheesh Mahal and Khas Mahal through intertwining paths. It is in these precincts that Shah Jahan spent eight years of his life (1658–1666) incarcerated by his son Aurangzeb. As a concession, a private mosque, Moti Masjid, was built here for his daily prayers (Siddiqi, 2008). The Musamman Burj or the octagonal tower sits elegantly on an elevated deck overlooking the river Yamuna (Fig. 4.32). The surface materiality is exquisite, fine

Figure 4.32 Musamman Burj and *jharokha* overlooks river Yamuna (Photo: Sakoon Singh)

marble embellished with precious *pietra dura*. The interiors of the pavilion are treated in the most breathtaking of design details with precious and semi-precious stone inlay work (Fig. 4.33). The center of this beautiful pavilion has a shallow water basin with a central fountain crafted in stone, which seems to melt into the floor level. The outer veranda facing river Yamuna is supported on faceted columns creating openings for views of the Taj Mahal. The inner structure is delicately detailed with floral motifs on columns, beams, capital and friezes (Siddiqi, 2008). Sheesh Mahal was built by Shah Jahan in 1630; the palace as the name suggests is studded with decorative mirror work on the walls and ceiling (Fig. 4.28; point 7).

Figure 4.33 Interiors of structures in and around Diwan-i-Khas, treated with fine inlay of semi-precious and precious stones

(Photo: Gurmeet Kaur)

Diwan-i-Khas (the Hall of Private Audience) comprises an outer columned hall and an inner closed hall, both connected by three multifoiled archways (Fig. 4.29; point 15). The outer hall measures 29.26 m × 10.1 m, and the inner one measures 12.20 m × 7.97 m (Siddiqi, 2008). The outer hall has a structural system of beautifully decorated marble columns that support the stone ceiling above and is further braced with multifoiled arches. The outer colonnades double up to create an elegant reading into the scale and proportion of Diwan-i-Khas. The aesthetic of ornamentation is exquisite with the generous use of marble inlays, embellishments with *pietra dura* and dado panels with relief carvings of floral motifs (Fig. 4.34). Inscriptions in Persian characters are inlaid in black marble on the south wall of the hall. The bejewelled Peacock throne was the most iconic attraction of this court which was later shifted to the Red Fort in Delhi. The inner hall is delicately screened by marble *jaalis* designed in floral patterns acting as a veil between the alcove and the raised seat for the emperor.

Figure: 4.34 *Pietra dura* and dado panels with relief carvings of floral motifs in Diwan-i-Khas

(Photo: Noor Dasmesh Singh)

4.6.5 Diwan-i-Am and Moti Masjid (Pearl Mosque)

The Diwan-i-Am (Fig. 4.28; point 11) was the hall of public audience where the emperor received audiences, reviewed troops and met people at large. It is a large open hall with colonnades and arches framing the hall in a symmetrical beauty that is awe inspiring. This hall consists of a large courtyard, and while the emperor took throne in the open hall, the general public congregated in the three cloisters. A raised throne was installed for the king, behind which a door led to the private ramparts of the palace. Diwan-i-Am is an enclosure which is more like a traditional veranda in its design principles being open on three sides. The fourth (eastern) side is solid with the *jharokha* that contained the celebrated throne *Takht-i-Murassa*, and the eastern wall has elaborate inlays done with *pietra dura*. The chamber with its richly decorated alcove is connected to Machhi Bhawan and onward to the rest of the palace. The royal ladies would use the maze of perforated marble screens to witness ceremonies. The Diwan-i-Am was built on a raised plinth 1.25 m high on a rectangular plan 61.77 m × 20.12 m. A total of 40 columns support the flat stone roofing of the hall, each one braced with engrailed arches (Siddiqi, 2008). On the outer side, the columns double up as twin sets of two, making the total number of columns in Diwan-i-Am 48. This colonnade structure was not originally intended to be used as Akbar's public audience hall but was later on adapted for this usage (Siddiqi, 2008; Verma, 1985).

The Moti Masjid (Fig. 4.28; point 14) was built by Shah Jahan towards the north of Diwan-i-Am and includes a deft use of marble, the surface of which shone, which earned it the epithet Pearl/"Moti". It was meant for use by members of the royal household.

The pastiche of influences that worked on the Agra city are thus visible in one variegated space of the Agra Fort (Koch, 1991). In its layered space it echoes all the influences the city of Agra witnessed in its winding history.

Chapter 5

Historical quarries of the Makrana Marble, the Vindhyan Sandstone and the Delhi Quartzite

An account

5.1 Introduction

The architectural heritage built and rebuilt over a period of time in the cities of Delhi and Agra led to enormous quarrying activity and the evolution of architectural and masonry techniques that were unique to these two historical cities of India. To throw light on the monuments and the stones used in them, it is imperative to take into account the numerous narratives written during the period when these heritage structures were erected. The four UNESCO Heritage Sites viz., Humayun's Tomb, Agra Fort, Red Fort Complex and Taj Mahal were erected during the reigns of emperors Akbar and Shah Jahan from the 16th to 17th century. The Mughal chroniclers, courtiers and historians of Emperor Akbar mostly recorded intricacies of the royal ceremonies, wars, religious discourses, tax systems, expansion of the governance etc., in the form of treatises and manuscripts. Akbar Nama, a set of three treatises, originally written in Persian by Abul Fazl Allami (courtier of Emperor Akbar), put forth the different facets of Akbar's empire and accounts of his predecessors. It is the most cited work of Abul Fazl in the contemporary times to comprehend the various dimensions of Akbar's vision, work and persona. The last treatise known as "Ain-I Akbari" has five volumes where "Ain" translates to "mode of governing" and "Ain-I Akbari" translates to "mode of governance by Akbar" (Blochmann, 1873). Careful perusal of the relevant volume I of Ain-I Akbari, having mention of monuments and materials used in them, reveals only sporadic and fragmentary accounts of these. This leaves a glaring lacuna in the field of architectural and building material details to construe (Blochmann, 1873; Phillotts, 1927; Nath, 1972). Abdul Hamid Lahauri, courtier of Emperor Shah Jahan, in his book titled *Padshahnama* recorded architectural and building material accounts of some of the edifices erected in Delhi and Agra by Shah

Jahan viz., Taj Mahal and later additions to the Agra Fort, and Delhi Red Fort. Nath (1972) mentions that the accounts of Lahauri specifically on the stones used in Taj Mahal are not accurate and he missed the significant details of architectural and building materials in his narrative. The other documented accounts on the UNESCO Heritage Sites and plethora of monuments from Agra and Delhi are from the Western/foreign travelers like Ralph Fitch, John Mildenhall, William Hawkins, William Finch, Peter Mundy, Nicholas Withington, Thomas Coryat and Edward Terry (Foster, 1921; Temple, 1914). The earlier-mentioned Mughal and foreign chroniclers, to some extent, recorded the UNESCO monuments of Delhi and Agra in their narratives, namely, *Ain-I-Akbari*, *The Travels of Peter Mundy in Europe and Asia*, *Early Travels in India*, *Shah Jahan Nama*, *Padshah Nama* etc., thus making it easier to interpret the source of the stones used in them. The accounts on Qutb Complex are mostly retrieved from the inscriptions on the stones used in the complex and narratives available through the writings of archaeologists and historians in the last 150 years.

The reference of red sandstone and their quarries, used extensively in the making of the Humayun's Tomb, Agra Fort and Fatehpur Sikri (also a UNESCO Cultural Heritage Site), Red Fort (Delhi) and scores of other edifices built during Akbar and Shah Jahan's authority, are referred to in the narratives of Abul Fazal, William Finch, Peter Mundy and Abdul Hamid Lahauri, to name a few. Some folklore suggests Akbar's fondness for the red sandstone as it symbolized power, which is evident from most of the monuments built in red sandstone during his reign. Likewise, Shah Jahan, who built Taj Mahal, was enthralled with the serene white marble which prominently adorns many magnificent edifices erected by him during his Golden reign in terms of Mughal architectural zenith in India. The letters written by Emperor Shah Jahan in relation to the supply of Sang-e-Marmar (white marble) to Mirza Raja Jaisingh of Ambar (Amer, modern Jaipur) are also referred to in sequel. The British officials, during their Raj in India, started the formal documentation of geographic, historical, economic and administrative affairs of India in the form of Imperial Gazetteers. *The Imperial Gazetteer of India* and *Imperial Gazetteer of India: Provincial Series* being the most reliable source of information. Thus, excerpts from numerous such Gazetteers chronicling the historical accounts on the monuments and building materials have been incorporated in this chapter.

The dominant stones/rocks used in plethora of architectural Heritage Sites of Delhi and Agra, including the UNESCO World Heritage Sites from these two places viz., Qutb Minar and its Monuments, the

Humayun's Tomb, the Red Fort Complex, the Taj Mahal and the Agra Fort, belong to the current states of Rajasthan, Uttar Pradesh and Delhi (Fig. 5.1). To further understand the chronicles and translated manuscripts beyond the excerpts given later, one can consult these Gazetteers, chronicles, manuscripts and books archived in the library of the Archaeological Survey of India (ASI) in Delhi and also in their online digital library. The accounts on quarries and stones used in the monuments

Figure 5.1 Map of India in 2019

(Source: traced from www.mapsofindia.com accessed on 26 September 2019)

quoted from the earlier narratives can be downloaded. We suggest curious readers to visit the on-line library sites like (https://dsal.uchicago.edu/) and (https://archive.org/) to get a better picture and grasp of the subject. These digital libraries give free access to most of the earlier chronicles and books. Wherever possible, these links are mentioned in the text for easy access to the readers who need a deeper intervention into the topic.

To begin with, one needs to understand the political division of India during the Mughal period, Colonial period and in the current scenario. The Mughal Empire, during Emperor Akbar's reign, mainly divided northern and central Indian subcontinent into 16 "Subas" (Provinces) which were further divided into numerous "Sarkars" (Districts). Figure 5.2

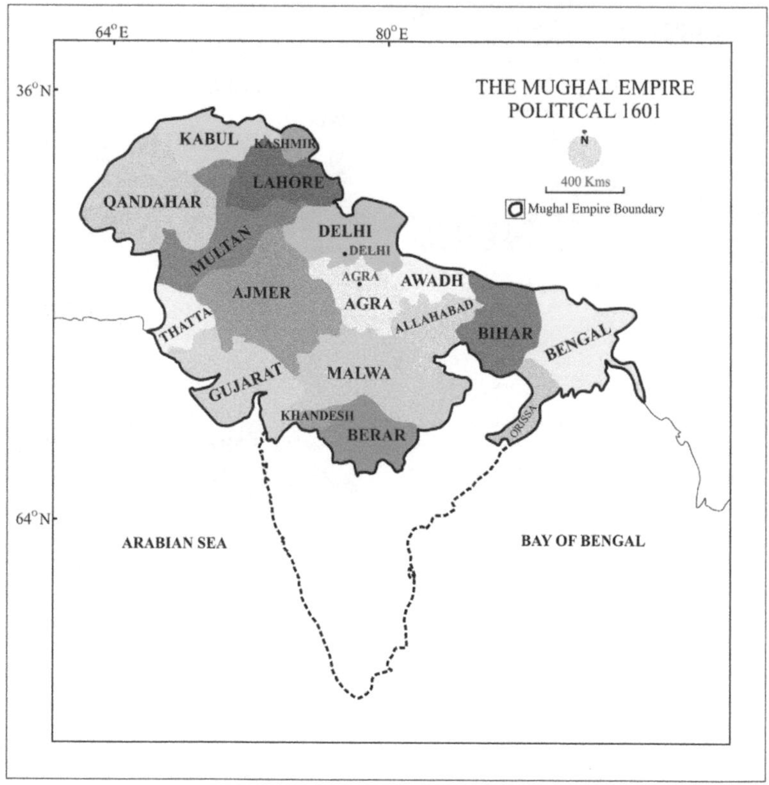

Figure 5.2 Political Map of Mughal Empire in 1601

(Source: after Habib, 1982)

is a political map of Akbar's empire during 1601 (Habib, 1982). Similarly, Figure 5.3 projects the political scenario of India during Colonial period of 1931 where the Indian subcontinent was divided into: British India, Territories permanently administered by the Indian Government and Indian states and Territories (https://dsal.uchicago.edu/reference/gaz_atlas_1931/pager.php?object=28). The current map of India, comprising 28 states and 9 Union Territories, is given for comparison with the political division of India during the Mughal and Colonial periods (Fig. 5.1). The political divisions during different periods make sense of the accounts written in the context of monuments, stones and the stone quarries during those times. The erstwhile boundaries/confines of Delhi and Agra during the Mughal, Colonial/British India and the current times, with reference to quarry sites of marble, sandstone and quartzites used in the UNESCO Heritage Sites of Delhi and Agra are shown in Figure 5.4, 5.5 and 5.6, respectively, to corroborate with the text below.

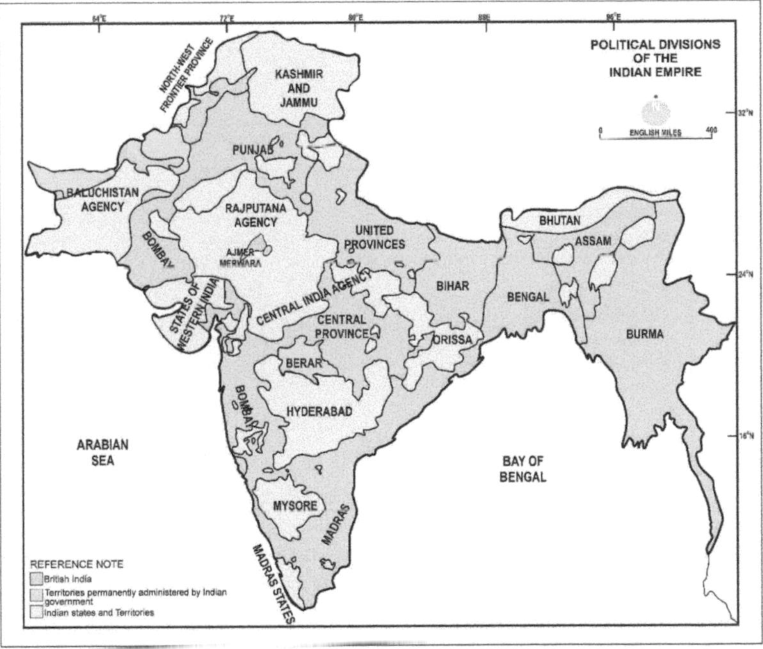

Figure 5.3 Political divisions of the Indian Empire in 1931

(Source: Frowde, 1908)

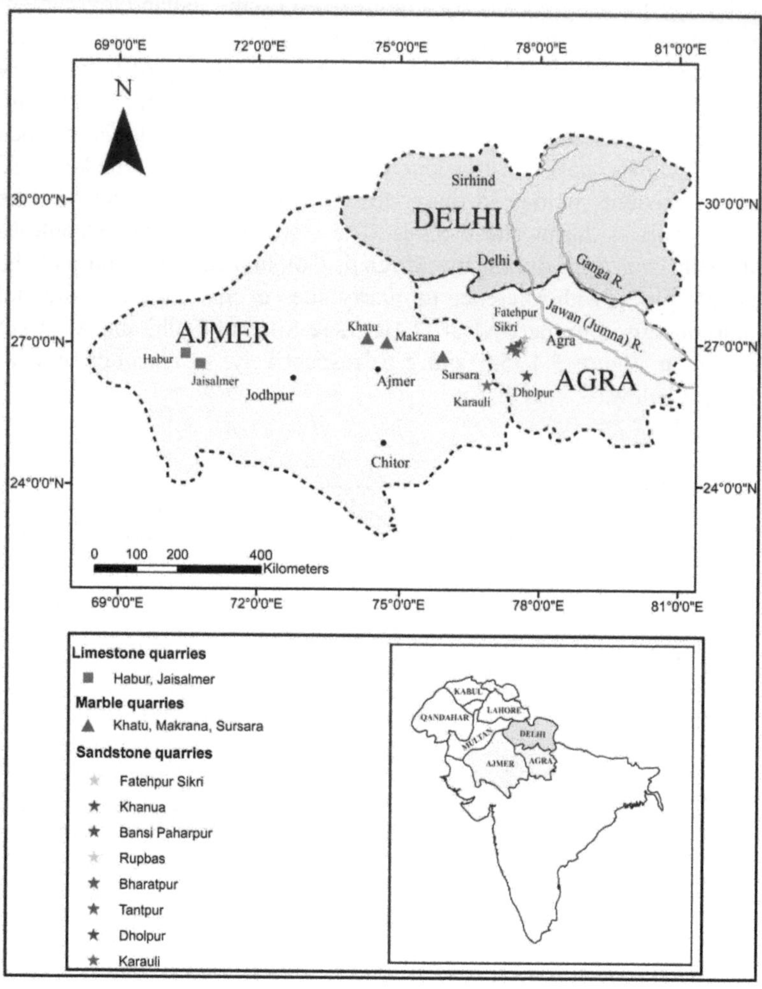

Figure 5.4 Map showing quarries of marble, sandstone and limestone in Ajmer and Agra "Subas" during Mughal period in 1601

(Source: base map Habib, 1982)

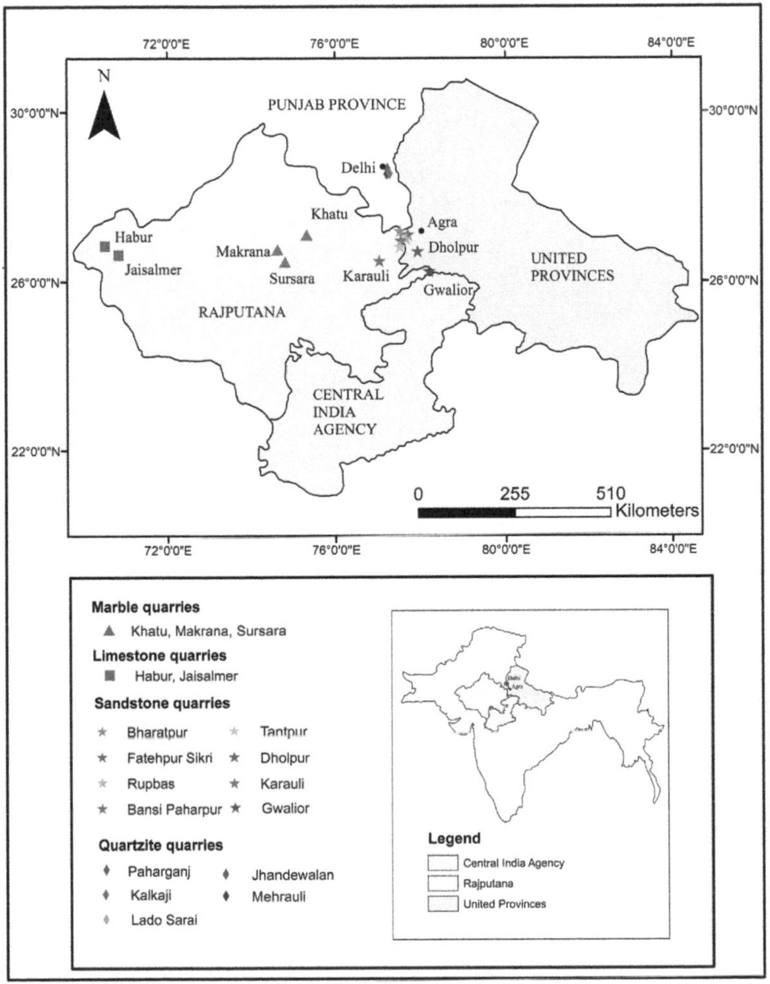

Figure 5.5 Map showing quarries of marble, sandstone, quartzite and limestone in Rajputana, United Provinces, Punjab Province and Central India Agency during British rule

(Source: base map Frowde, 1908)

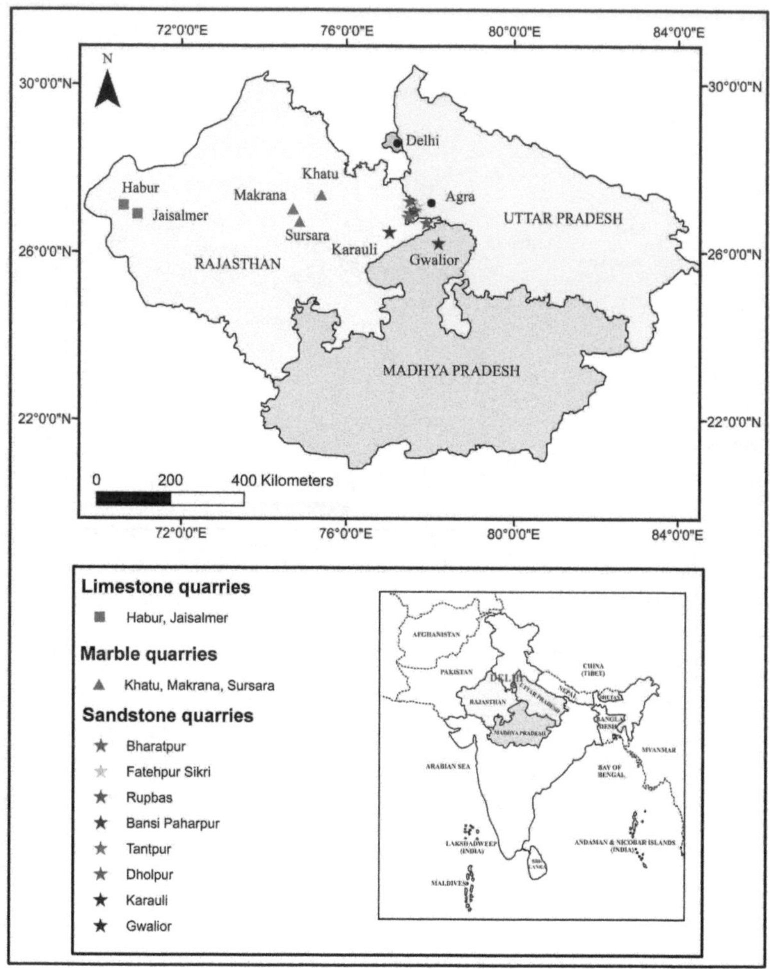

Figure 5.6 Map showing quarries (some of which are historical quarries) of marble, sandstone and limestone in Rajasthan, Uttar Pradesh and Madhya Pradesh in contemporary India

(Source: base map from Google Earth)

5.2 Account of historical quarries used in the UNESCO Heritage Sites of Delhi and Agra

The Agra Fort was commissioned under the patronage of Emperor Akbar and subsequent additions were made by Shah Jahan into the fort complex. The use of red sandstone in the monuments built in Agra and Fatehpur Sikri (a city southwest of Agra) finds mention in the narratives of Abul Fazl in Ain-I Akbari (Volume I). The forts, palaces, victory gates, *minars* (towers), tombs, *kos minars* (mile stones), *baolis* (tanks) along with many other edifices, were commissioned in Agra and Fatehpur Sikri using well-known red sandstone and its varieties from the nearby Vindhyan hillocks exposed in and around Fatehpur Sikri. The excerpts from the chronicles of Abul Fazl (Blochmann and Phillotts, 1927) indicate the abundance of edifices and sandstone exposures in the vicinity of Fatehpur Sikri and Agra.

Original "IMPERIAL FIRMANS" (Imperial orders) of Emperor Shah Jahan on supply of marble from Makrana for construction of Taj Mahal are available in Bikaner Archives, Rajasthan. The Firmans given below have been procured from the book authored by Nath (1985). The translated Firmans clearly indicate the source of marble for the Taj Mahal was Makrana Quarries of Ajmer province of the Mughal Empire. The marble quarrying subsequently continued from the Rajputana state of Colonial India and to date is being carried out in Makrana in present the state of Rajasthan. Readers may refer to the original Firmans currently preserved in Bikaner Archives, Rajasthan; their photo productions are archived in several books (Nath, 1985; Begley and Desai, 1989). In the three Firmans, sent by Emperor Shah Jahan to Mirza Raja Jaisingh of Ambar (Amer, modern Jaipur), reference to construction of Taj Mahal vis-à-vis marble and skilled labor is explicit.

Excerpts on building material from Mughal province "Ajmer", British India "Rajputana" and current day "Rajasthan" state of India compiled from various Chronicles, Gazetteers are given below:

A'in (mode of governing) 85

ON BUILDINGS

Hence His Majesty plans splendid edifices, and dresses the work of his mind and heart in the garment of stone and clay. Thus mighty fortresses have been raised, which protect the timid, frighten the rebellious, and please the obedient. Delightful villas, and imposing towers have also been built.

Source: Blochmann (1873, p. 232), Phillott (1927, p. 233)
Web Source: https://archive.org/details/in.ernet.dli.2015.279471/page/n279, accessed September 28 2019

A'in (mode of governing) 86

PRICES OF BUILDING MATERIAL ETC

Red sandstone costs 3 dams per man. It is obtainable in the hills of Fatehpur Sikri, His Majesty's residence, and may be broken from the rocks at any length or breadth. Clever workmen chisel it so skillfully, as no turner could do with wood; and their works vie with the picture book of Mani (the great Painter of the Sassanides). Pieces of red sandstone (sang-i-Gulûla), broken from the rocks in any shape.

Source: Blochmann (1873, p. 223), Phillott (1927, p.233)
Web Source: https://archive.org/details/in.ernet.dli.2015.279471/page/n280, accessed September 28 2019

About 3 Course off lies Rupbaz [Rupbas] where are the quarries of those redd stones, which supplye all their parts for the principall buildings, as the Castle of Agra, this place, Greatmens howses, Tombes, etts.

Source: Temple (1914, p. 231)
Web Source: https://archive.org/details/travelsofpetermu02mund/page/n343, accessed September 28 2019

It is remarkable that the quarries of India, specially neere Fettipore (whenee they are carryed farre) are of such nature that they may be cleft like logges and sawne like plancks to seele chambers and cover houses of a great length and breadth. From this monument is said to bee a way under ground to Dely castle. (sic)

Source: Foster (1921, p. 157)
Web Source: https://archive.org/details/in.gov.ignca.3220/page/n187, accessed September 28 2019

Bansi Paharpur red sandstone quarries

Sandstone is plentiful almost everywhere, varying greatly in texture and colour. The most famous sandstone quarries are at Bansi Paharpur in Bharatpur State; they have furnished materials for the most celebrated monuments of the Mughal dynasty at Agra, Delhi, and Fatehpur Sikri, as well as for the beautiful palaces at Dig. There are two varieties of this stone: namely, a very fine-grained yellowish white which; and a dark red, speckled with yellow or white spots. The quarries give employment to 450 laborers, and the outturn is about 14,000 tons a year.

Source: Erskine (1908, pp. 53–54)
Web Source: https://archive.org/details/in.ernet.dli.2015.207013/page/n87, accessed September 30 2019

Bansi Paharpur sandstone quarries

The famous sandstone quarries at Bansi Paharpur furnished materials for the most celebrated monuments of the Mughal dynasty at Agra, Delhi, and Fatehpur Sikri, as well as for the beautiful palaces at Dig. The stone is of two varieties: namely, dark red, generally speckled with yellowish white spots or patches; and a yellowish white, homogeneous in colour and texture, and very fine-grained. The red variety is inferior for architectural purposes to the white, but is remarkable for perfect parallel lamination; and, as it readily splits into suitable flags, it is much used for roofs and floors. The annual outturn is about 14,000 tons, of which about two-thirds are sold to the public on payment of royalty, and the balance is utilized for State works. These quarries give employment to some 450 labourers who are mostly Ujhas (or carpenters) residing in the neighbourhood, and whose monthly earnings average Rs.6 to Rs.10 per head.

Source: Erskine (1908, p. 330)
Web Source: https://archive.org/details/in.ernet.dli.2015.207013/page/n363, accessed September 30 2019

Dholpur sandstone quarries

The red sandstone of Dholpur is most valuable for building purposes; fine-grained and easily worked, it hardens by exposure, and does not deteriorate by lamination. The principal quarry are at Narpura, 4 miles north-west of the capital, with which they are connected by a railway siding, and near Bari; they are worked on the petty contract system, and in 1900–1 yielded a net profit of Rs. 13,300, which had increased to Rs. 21,300 in 1904–5.

Source: Erskine (1908, pp. 345–346)
Web Source: https://archive.org/details/in.ernet.dli.2015.207013/page/n379, accessed September 30 2019

Karauli sandstone quarries

Red sandstone abounds throughout the greater portion of the State and in parts, especially near the capital, white sandstone blends with it. Other varieties of a bluish and yellow colour are also found, the former near Machilpur, and the latter in the south and west.

Source: Erskine (1908, p. 358)
Web Source: https://archive.org/details/in.ernet.dli.2015.207013/page/n391, accessed September 30 2019

Rupbas sandstone quarries

Headquarters of a tehsil of the same name in the State of Bharatpur, Rajputana, situated in 26^{0} 59'N and 77^{0} 39' E about 19 miles south-by-south-east of Bharatpur city. . . . In the vicinity of Rupbas are some enormous stone obelisks and images. The oldest is a sleeping figure of Baldeo cut in the rock, 22 ½ feet long, with a seven serpent-hooded canopy and an inscription dated A.D. 1609. About 8 miles to the south-west are the famous sandstone quarries of Bansi Paharpur, which have supplied material for the beautiful palaces at Dig and for many of the buildings at Agra and Fatehpur Sikri.

Source: Erskine (1908, pp. 339–340)
Web Source: https://archive.org/details/in.ernet.dli.2015.207013/page/n373, accessed September 30 2019

IMPERIAL FIRMANS

THE FIRST TRANSLATED FIRMAN/IMPERIAL ORDER DATED 21ST JANUARY 1632 READS AS FOLLOWS:-

As a great number of carts is required for transportation of marble needed for constructing buildings (at the Capital), a firman was previously sent (to you to procure them). It is, again, desired of you that as many carts on hire be arranged as possible at the earliest time, as has already been written to you, and be sent to Makrana for expediting the transport of marble to the Capital (Agra). Every assistance be given to Allãh-dãd (Khan) who has been doputed to arrange transportation of marble to Akbarabad (Agra).

Account (of the expenditure on carts) along with the previous account of amounts allocated for the purchase of marble be submitted (to the Mutasaddî in-charge of payments).

Source: Nath (1985, Appendix IV, p. 64)

THE SECOND TRANSLATED FIRMAN DATED 9TH SEPTEMBER 1632 READS AS FOLLOWS:-

Malük Shäh has been deputed to Ambar (Amer) to bring marble from the new mines (of Makrana). It is commanded that carts on hire be arranged for transportation of marble, and Maluk Shah be assisted to purchase as much marble as he may desire to have. The purchase price of marble and cartage shall be paid by him from the Royal Treasury. Every other assistance be given to him to procure and bring marble and sculptors to the Capital expeditiously.

Source: Nath (1985, Appendix IV, p. 65)

THE THIRD TRANSLATED FIRMAN DATED 21ST JUNE 1637 READS AS FOLLOWS:-

We hear that your men retain the stone-cutters (sang-tarãsh) of the region at Ambar (Amer) and Rajnagar. This creates shortage of stone-cutters (miners) at Makrana and the work (of procuring marble) suffers. Hence, it is desired of you that no stone cutter be detained at Ambar and Rajnagar, and all available miners be sent to the Mutasaddis of Makrana (so that marble be procured in large quantities to be dispatched to the Capital for building purpose).

Source: Nath (1985, Appendix IV, pp. 65–66)

The 13th March 1632/3. Att 4 in the morninge wee stayed heere amonge the Hills [Aravalli Mts.], (Setila [now Satpura], 6 course), our Cammells and Oxen not being able to followe Backur Ckaun, who went [on] to Adgemeere [Ajmer]; our waie stonie lookeinge like Marble.

Some 7 Course off is Nurnoulee, from whence are brought all your Marble stones, wherewith the kinge is supplyed for his buildinges, there being noe lesse then 500 Carts Comeing and goeinge in its carriage [i.e., for its transport].

Wee past by Kissungurre [Kishangarh], a Castle with a Cittie under it, Hard by a learge Tanck [Gund Talao].

Source: Temple (1914, p. 241)

Web Source: https://archive.org/details/travelsofpetermu02mund/page/240, accessed September 28 2019

Makrana Marble quarries

Village in the Parbatsar district of the State of Jodhpur, Rajputana situated in 27⁰ 3' N and 74⁰ 44' E, on the Jodhpur-Bikaner Railway. Population (1901), 5, 157. The village derives its importance from its marble quarries, which have been noted for centuries, and from which the material used in the construction of the Taj Mahal at Agra was obtained. It has been proposed to use this marble for the Victoria Memorial Hall at Calcutta. The quarries vary in depth from 30 to 75 feet and the yearly out-turn averages about 900 or 1000 tons. The marble is excavated by blasting, and is then cut into required sizes by means of steel saws. The chips and dust left behind after the blocks have been hauled to the surface are burnt into lime and used for the finer kinds of plastering. There are now twenty-six quarries being worked, which give employment to about 100 labourers daily, mostly of the Silawat caste of Muhammadans.

Source: Erskine (1908, p. 199)

Web Source: https://archive.org/details/in.ernet.dli.2015.207013/page/n233, accessed September 30 2019

Limestone is abundant in several parts, and is used both for building and for burning into lime. Two local forms- of it stand pre-eminent among the ornamental stones of India for their beauty; namely, the Raialo group, quarried at Raialo (Raiala) in Jaipur, at Jhiri in Alwar, and at Makrana in Jodhpur; and the Jaisalmer limestone. The former is a fine grained crystalline marble, the best being pure white in colour, while others are grey, pink, or variegated. The famous Taj at Agra was built mainly of white Makrana marble, and it is proposed to use the same stone in the construction of the Victoria Memorial Hall at Calcutta. The Jaisalmer variety is of far later geological age; it is even-grained, compact, of a buff or light brown colour, and is admirably adapted for fine carving. It takes a fair polish, and was at one time used for lithographic blocks.

Source: Erskine (1908, p. 53)

Web Source: https://archive.org/details/in.ernet.dli.2015.207013/page/n87, accessed September 30 2019

Delhi quartzite and Alwar slate quarries

The rocks of Rajputana are rich in good building materials. The ordinary quartzite of the Aravallis is well adapted for many purposes; the more schistose beds are employed as flagstones or for roofing, and slates are found in the Alwar and Bundi hills.

Source: Erskine (1908, p. 53)

Web Source: https://archive.org/details/in.ernet.dli.2015.207013/page/n87, accessed September 30 2019

Jaisalmer and Habur limestone quarries

> There are several quarries of limestone near the capital; the stone produced is very fine, even-grained, and compact, of a buff or light-brown colour, and admirably adapted for carving. It takes a fair polish, and was at one time used for lithographic blocks. Another variety of yellow limestone is found at the village of Habur, 28 miles north-west of the capital; large quantities of an iron ore resembling red ochre are blended with it.

Source: Erskine (1908, p. 212)

Web Source: https://archive.org/details/in.ernet.dli.2015.207013/page/n245, accessed September 30 2019

Careful scrutiny of the above excerpts makes obvious the predominant use of stones during the construction of most of the magnificent edifices during the 16th to 17th century. The dominant stones viz., marble, sandstone and quartzite used in the UNESCO Heritage Sites of Delhi and Agra were sourced from northern Aravalli Mountain Belt and Vindhyan Basin (Fig. 5.7). The red sandstone and its other varieties in the vicinity of Fatehpur Sikri were present in abundance and belonged to the present day Western Sector of the Vindhyan Basin. That still holds true as red sandstone is being quarried from the historical quarries of red sandstone in large quantities. The red sandstone and its varieties belong to the Upper Bhander Group of Vindhyan Supergroup. The most popular sandstone quarries which provided sandstones for the mentioned UNESCO monuments were Bansi Paharpur, Rupwas, Dholpur, Tantpur and Fatehpur Sikri.

Large quantities of marble were quarried from Makrana to ensure uninterrupted supply of the marble for the construction of the Taj Mahal. The marble from Makrana was referred to as '*Sang-e-Marmar*' in Persian which translates to flawless white marble. The Makrana marble was transported from Makrana in Ajmer province to Agra province through hundreds of carts in different phases, as it took almost 17 long years to build it. Numerous stone cutters were deployed in the quarries to accomplish the then ardent task.

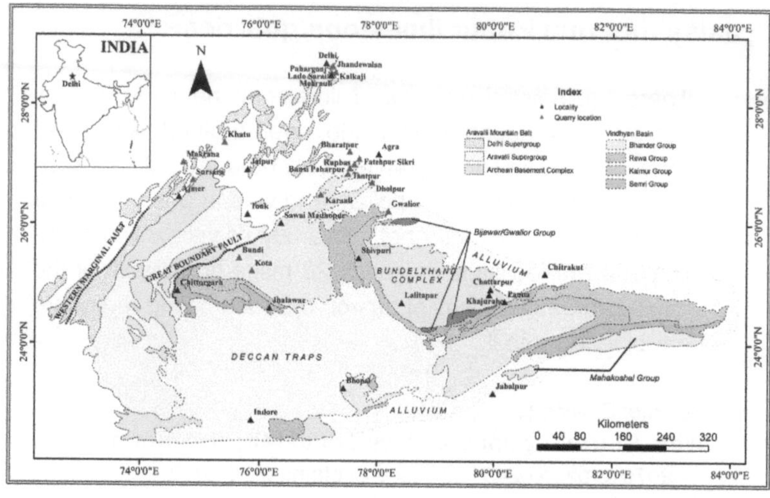

Figure 5.7 Geological map of Aravalli Mountain Belt and Vindhyan Basin with quarry locations of marble, sandstone, quartzite (now banned) and limestone

(Base map sources: Prasad, 1984; Deb *et al.*, 2001; Cavallo and Pandit, 2008; Malone *et al.*, 2008)

5.3 Current status of Makrana Marble deposits and quarries

The Makrana marble has been quarried in the vicinity of the Makrana town of Rajasthan for more than four centuries which resulted in the total landscape changeover of the area. Makrana, world famous for marble quarries, was instrumental in meeting the marble needs of numerous edifices built during the reign of Emperor Shah Jahan. Because the quarrying has never ceased in Makrana area since its inception, it has led to deeper pits in the region and the quarrying has become unsafe at many places (Fig. 5.8). At the same time, these quarries are a major source of income for the unskilled, semi-skilled and skilled miners living in the town and neighboring areas (Garg *et al.*, 2019). Roughly 800 quarries are operational over an area of approximately 30 sq km to the northwest, west and southwest of Makrana town. The Makrana quarries operational in various bands of marble in the Ras Formation of Kumbhalgarh Group of the SDFB are shown on Figure 5.9 and 5.10 (Table 5.1 and 5.2; Bhadra *et al.*, 2007). Garg *et al.* (2019) based on counting on the satellite image procured from

Figure 5.8 Makrana marble quarries in the vicinity of Makrana town (Photos: Anuvinder Kaur)

European Space Agency for the year October 2017 (Fig. 5.11) have reported around 745 functional quarries in contrast to around 550 functional quarries reported by Bhadra *et al.* (2007) based on counting on an Indian Remote Sensing Satellite image procured in the year 2003 (Table 5.2; Fig. 5.10). Amongst all the marble bearing bands in the Makrana region (Table 5.1), Devi-Gunawati range is known for the best quality of Makrana white marble from quarries of Chousira, Ulodi and Pahad Kuan (Fig. 5.9; Garg *et al.*, 2019 and references cited therein).

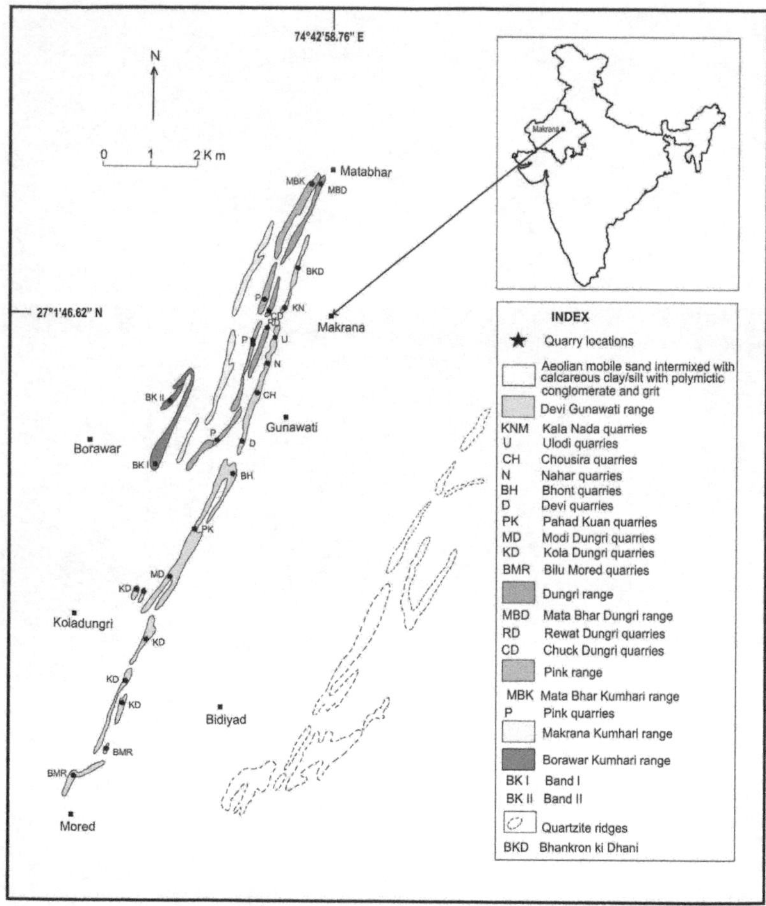

Figure 5.9 Geological map with Makrana marble quarries, Nagaur district, Rajasthan

(Source: modified after Natani and Raghav, 2003)

Table 5.1 Description of Makrana marble bands in Kumbhalgarh Group of SDFB

Name of the band	*Length of the band*	*Width*	*Important quarries*	*Variety of marble*
Devi-Gunawati Band	13 kms	90–150 mts	Kala Nada quarries Ulodi quarries Chousira quarries Nahar quarries Bhont quarries Devi quarries Pahad Kuan quarries Modi Dungri quarries Kola Dungri quarries Bilu Mored quarries	Pink, white, multicolored and pure white (Makrana white marble)
Dungri Band	4 kms	60–80 mts	Mata Bhar dungri quarries Rewat dungri quarries Chuck dungri quarries	Super white (Makrana white marble) and variegated white
Pink Banc	1.7 kms	35–70 mts	Mata Bhar Kumhari quarries Pink quarries	Light pink (pink variety)
Makrana Kumhari Band	3.5 kms	40–50 mts		Greyish-bluish white (Dungri variety)
Borawar Kumhari Band I & II	2.5 kms	30–40 mts	Band I Band II	Greyish white (Dungri variety); dark grey and greyish green (Kumhari variety)

(Source: Natani, 2000; Natani and Raghav, 2003; Bhadra *et al.*, 2007)

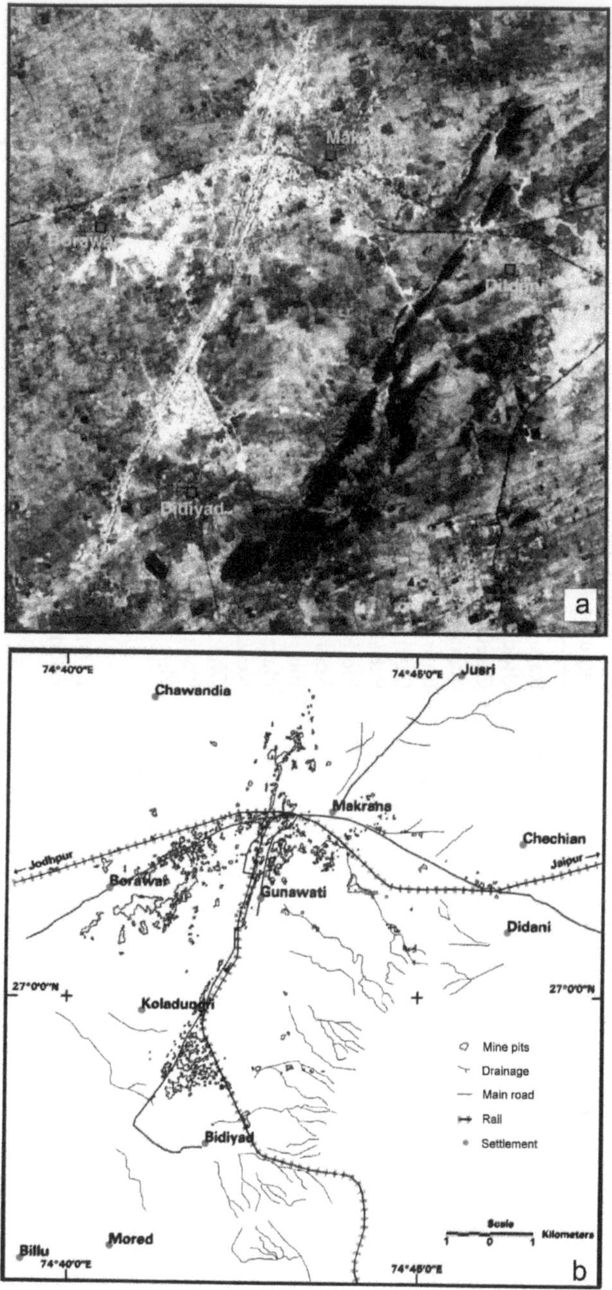

Figure 5.10 (a) IRS image of Makrana region (2003); (b) Digitized Makrana marble quarry sites in the earlier satellite image

(Source: after Bhadra *et al*., 2007)

Table 5.2 Quarries from Makrana marble ranges deduced from satellite image of 2003

Quarry area (m²)	*No. of quarry pits*	*Quarry size*
< 5,000	440	495 Small
5,000 to 10,000	55	
10,000 to 20,000	35	56 Medium
20,000 to 30,000	8	
30,000 to 40,000	5	
40,000 to 50,000	3	
50,000 to 60,000	3	
60,000 to 70,000	1	
70,000 to 80,000	0	
80,000 to 90,000	1	
90,000 to 1,00,000	1	1 Large

(Source: Bhadra *et al.*, 2007)

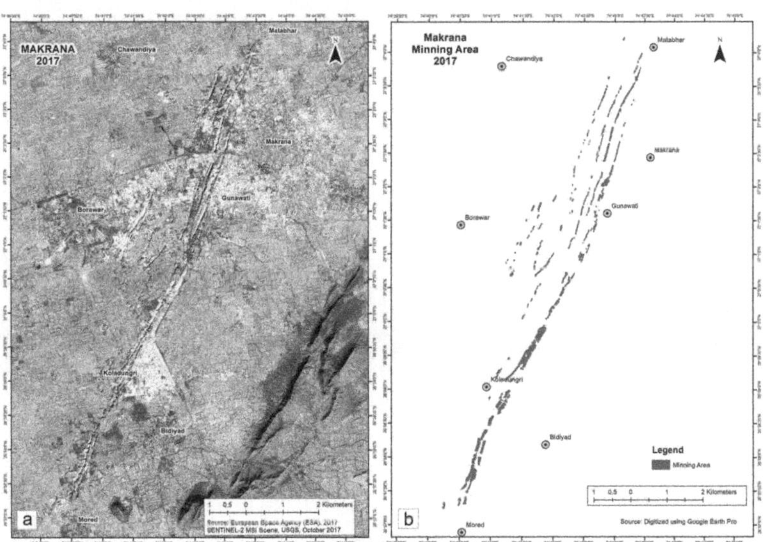

Figure 5.11 (a) European Space Agency Satellite image of Makrana region for 2017; (b) Digitized Makrana marble quarry sites on the earlier satellite image for 2017

(Sources: https://earthexplorer.usgs.gov/, accessed June 30 2017; after Garg *et al.*, 2019)

5.4 Commercial (Trade) varieties of the Makrana Marble

The Makrana Marble is traded as variants based on its color and textural patterns (Fig. 5.12; www.rkmarblesindia.com/indian-marbles/makrana-marbles). The six popular commercial varieties are: Albeta Marble: commonly contains streaks viz., the one with black streaks is commonly referred to as (i) Albeta Marble (Fig. 5.12a) and the one with brown linear patterns is known as (ii) Albeta Brown Marble (Fig. 5.12b); (iii) Kumhari Marble: white marble with grey brown linear patterns (Fig. 5.12c); (iv) Dungri Marble: streaks of grey/brown in the white marble (Fig. 5.12d); (v) Makrana Pink Marble: pink marble with grey streaks (Fig. 5.12e); and (vi) Makrana White Marble: it is a flawless white marble (Fig. 5.12f). The

Figure 5.12 Polished slabs of commercial/trade varieties of Makrana Marble: (a and b) Albeta Marble; (c) Kumhari Marble; (d) Dungri Marble; (e) Makrana Pink Marble and (f) Makrana White Marble

(Source: after Garg *et al.*, 2019)

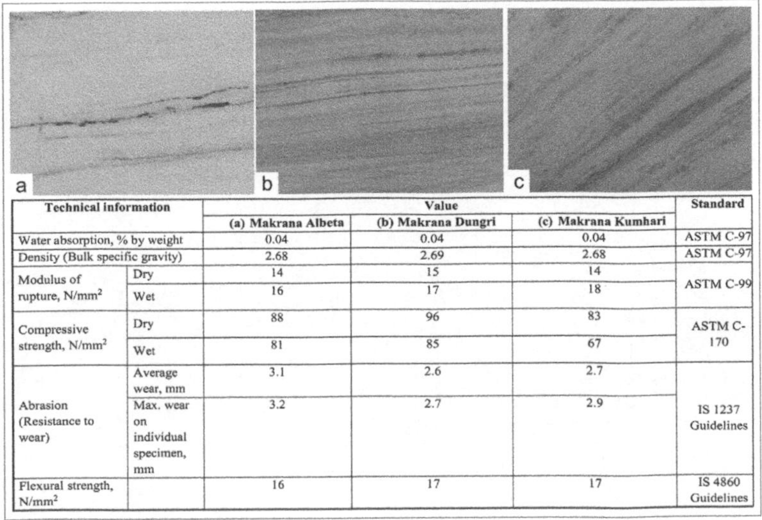

Technical information		Value			Standard
		(a) Makrana Albeta	(b) Makrana Dungri	(c) Makrana Kumhari	
Water absorption, % by weight		0.04	0.04	0.04	ASTM C-97
Density (Bulk specific gravity)		2.68	2.69	2.68	ASTM C-97
Modulus of rupture, N/mm^2	Dry	14	15	14	ASTM C-99
	Wet	16	17	18	
Compressive strength, N/mm^2	Dry	88	96	83	ASTM C-170
	Wet	81	85	67	
Abrasion (Resistance to wear)	Average wear, mm	3.1	2.6	2.7	IS 1237 Guidelines
	Max. wear on individual specimen, mm	3.2	2.7	2.9	
Flexural strength, N/mm^2		16	17	17	IS 4860 Guidelines

Figure 5.13 Technical properties of Makrana Marble (a) Albeta, (b) Dungri and (c) Kumhari varieties

(Source: Center for Development of Stones (CDOS), Jaipur)

Makrana White Marble is referred to as Sang-e-Marmar in Persian and was the most prized stone preferred by Emperor Shah Jahan for construction of many monuments in the 17th century. In contemporary times as well, it is the most sought after and expensive variety which is preferably used for sculpturing and is mostly exported for the same reason (personal communication with the marble workshop owners of Makrana). The Makrana Marble is found to be on a par or even better than most of the European marbles (Dube, 2008). Figure 5.13 provides physico-mechanical properties of three varieties of Makrana Marble.

5.5 Current status of Vindhyan Sandstone deposits of eastern Rajasthan

The Vindhyan Supergroup rocks are found in Western Sector and Eastern Sector of Vindhyan Basin (Fig. 5.7). These two sectors are dotted with sandstone quarries mostly in the Upper Bhander Group. The Rajasthan State Mines and Geology Department reports the occurrence of splittable sandstone from many districts and localities viz. Bharatpur district: Rupwas, Bansi Paharpur and Bayana (Fig. 5.14);

Figure 5.14 Vindhyan Sandstone quarries (a) Bansi Paharpur and (b) Rupbas (Photos: Gurmeet Kaur and Jaspreet Saini)

Table 5.3 Sandstone quarry localities in Karauli, Bharatpur and Dholpur districts

S. No	*Districts*	*Localities*	*Nature of Vindhyan Sandstone*
1	**Karauli District**	Makanpur-Batda, Garhi Ka Gaon, Bhankri in Mandrayal Tehsil, Lotda, Mahua Khera, Behrai in Masalpur Tehsil, Atewa-Kalyani-Mamchari in Karauli Tehsil and Bahadurpur of Sapotra Tehsil	Splittable/blockable variety mostly used as dimension stone
2	**Bharatpur District**	Roopwas Tehsil at Sirondh, Banshi Paharpur, Rajpura, Chaikoraand at Churari Dang, Basai Khori, Mahalpur Chura, Kharga Ka Nagla	Splittable/blockable variety; spotted and banded, both variety
3	**Dholpur District**	Bari, Sar Mathura and Baseri Tehsils. The main mining areas are Chilachondh, Naksoda, Sanaura of Bari Tehsil, Kachhpura, Tarwa, Math Pipraundh, Barauli, Liloti, Khurdia, Amanpura, Badagaon, Khushalpura, Chandpura, Kota, Sar Mathura. Kanchanpura and Maharpur in Sar Mathura Tehsil and Vijaypura, Bansrai, Tajpura, Nadaripur, Tilua in Baseri Tehsil.	Splittable/blockable variety; beige and red variety

(Source: Ra asthan State Mines Department; Kaur *et al.*, 2019b)

Karauli district: Hindaun, Mandrayal and Masalpur; Dholpur district: Sawai Madhopur, Dholpur, Kota, Badi, Basedi and Dhanodi; Kota district: Digod, Ladpura, Pipalda and Sangod; Bundi district: Khera, Rampuriya, Budhpura, Guda, Palkan, Nareli, Chhat Ka Khera and Thadi; Baran district: Attru and Mangrol; Bhilwara district: Bijoliya, Kasya and Hemniwas; Jhalawar district: Mandalgarh, Kansiya, Bania ka Talab, Aklrea, Khanpur and Jhalra Patan; Tonk and Chittorgarh in eastern Rajasthan (Kaur *et al.*, 2019b). Most of these quarrying localities (Table 5.3) are marked in Figure 5.15. The Bharatpur, Karauli and Dholpur districts host historical quarries of Rupbas, Bansi Paharpur, Tantpur and Dholpur, which were sources for sandstone to most of the monuments built from the 16th to 19th century during the Mughal period in India, including Agra Fort, Taj Mahal, Red Fort and Humayun's Tomb. Mention of these historical quarries in narratives of Mughal courtiers and foreign travelers is explicit, and the excerpts in the preceding section of this chapter leave no room for any doubt on the source of sandstone used in the UNESCO designated Cultural Heritage Sites of Agra Fort, Taj Mahal, Red Fort and Humayun's Tomb complex.

5.6 Commercial (Trade) varieties of the Vindhyan Sandstone

The Vindhyan Sandstones are traded by the commercial names such as Karauli Red, Dholpur Beige, Dholpur Red, Dholpur Pink, Bansi Red, Bansi Pink, Agra Red, Rupwas Red, Rupwas spotted etc., based on the locality names of their occurrence and their colors (Figs. 5.15 and 5.16). The Bansi Paharpur hosts some of the best quality sandstone in the region. The sandstone in Bansi Paharpur is thickly bedded, fine- to medium-grained and mainly pinkish and reddish pink in color. Manual to semi-mechanized and mechanized methods of quarrying are generally adopted in these areas. The physico-mechanical properties of three varieties of Vindhyan Sandstone are furnished in Figure 5.16.

5.7 Delhi Quartzite and quarries

Delhi city has sporadic exposures of the Delhi quartzite rocks of the Aravalli Mountain Belt surrounded by Newer and Older alluvium. Central and south Delhi and adjoining parts of Haryana have good

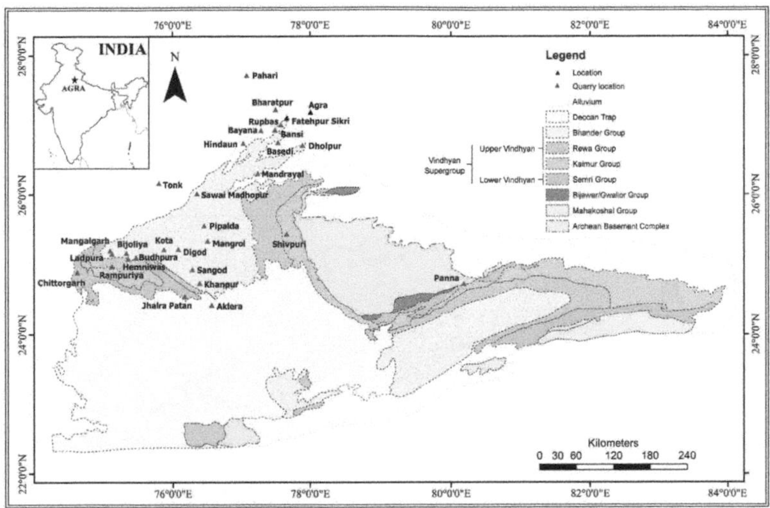

Figure 5.15 Geological map of Vindhyan Basin with sandstone quarries of western sector marked on it

(Source: base map adapted from Prasad, 1984 ; Malone *et al.*, 2008)

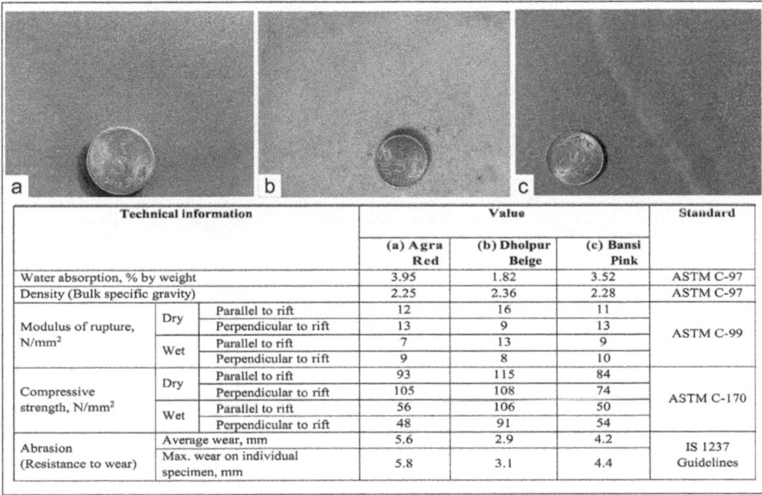

Technical information			Value			Standard
			(a) Agra Red	(b) Dholpur Beige	(c) Bansi Pink	
Water absorption, % by weight			3.95	1.82	3.52	ASTM C-97
Density (Bulk specific gravity)			2.25	2.36	2.28	ASTM C-97
Modulus of rupture, N/mm²	Dry	Parallel to rift	12	16	11	ASTM C-99
		Perpendicular to rift	13	9	13	
	Wet	Parallel to rift	7	13	9	
		Perpendicular to rift	9	8	10	
Compressive strength, N/mm²	Dry	Parallel to rift	93	115	84	ASTM C-170
		Perpendicular to rift	105	108	74	
	Wet	Parallel to rift	56	106	50	
		Perpendicular to rift	48	91	54	
Abrasion (Resistance to wear)	Average wear, mm		5.6	2.9	4.2	IS 1237 Guidelines
	Max. wear on individual specimen, mm		5.8	3.1	4.4	

Figure 5.16 Technical properties of (a) red, (b) pink and (c) beige varieties of Bhander sandstone of Rajasthan sector of Vindhyan Basin

(Source: CDOS, Jaipur)

exposures of the Delhi quartzites. The Delhi quartzite can be seen along the Delhi ridge which is NNE-SSW trending, and good exposures are seen at Mehrauli (south Delhi) and Surajkund (Haryana). The quartzite has been extensively used in the monuments of Delhi as

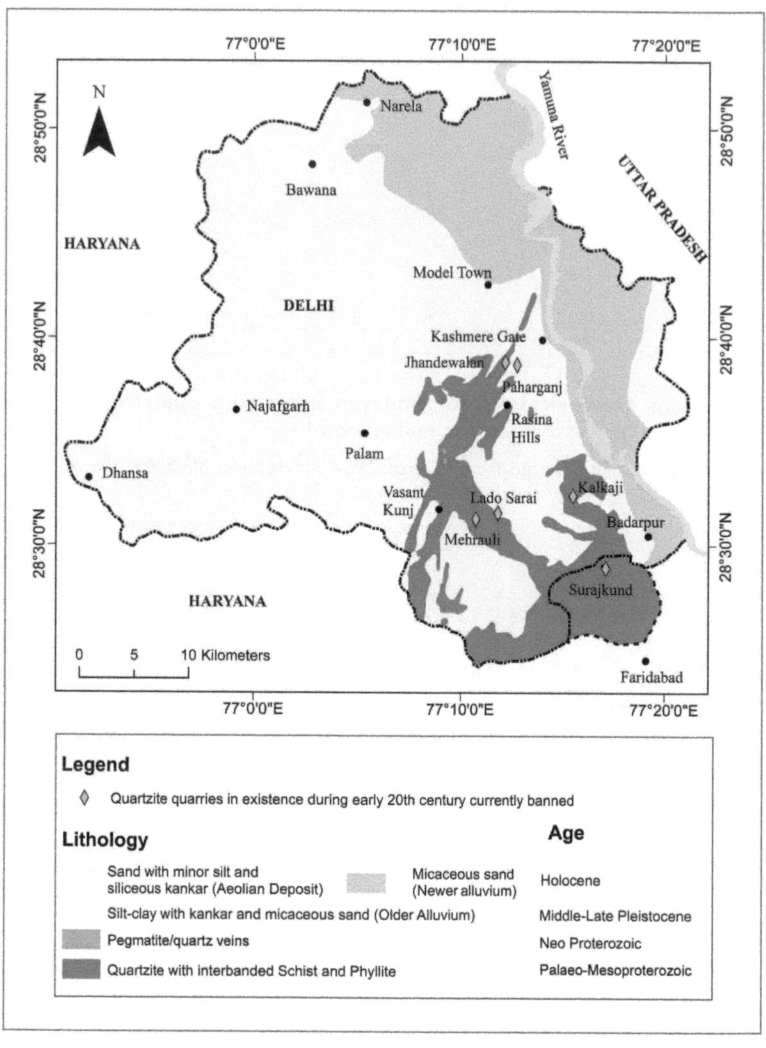

Figure 5.17 Geological map of Delhi and part of Haryana with erstwhile quartzite quarries marked on it

(Source: modified after Thussu, 2006)

a rubble masonry stone which was used to build the walls of various forts, complexes etc. The city of Delhi had many quartzite quarries inside the premises of city and in the adjoining areas covering parts of Haryana viz., Surajkund and Faridabad (Fig. 5.17). The physico-mechanical properties of the Delhi Quartzite are shown in Figure 5.18. The easy and local availability besides its competent nature made it a favorite building material for many architectural Heritage Sites of city of Delhi. The quartzites were extensively quarried at Paharganj, Jhandewalan, Kalkaji, Rohtak Road, Mehrauli, Rathia Lado Sarai and other places (Fig. 5.7; Chopra, 1976). The quarrying of Delhi quartzite has been lately banned inside and outside Delhi by the Supreme Court to arrest further environment degradation. The abandoned quarries have not been properly reclaimed and at some places these have affected the landscape of the city with quarry depressions filled with water and vegetation.

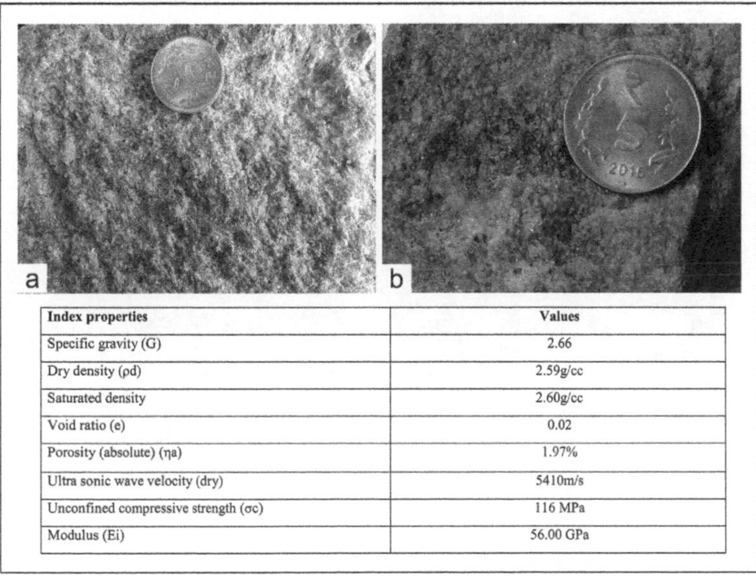

Index properties	Values
Specific gravity (G)	2.66
Dry density (ρd)	2.59g/cc
Saturated density	2.60g/cc
Void ratio (e)	0.02
Porosity (absolute) (ηa)	1.97%
Ultra sonic wave velocity (dry)	5410m/s
Unconfined compressive strength (σc)	116 MPa
Modulus (Ei)	56.00 GPa

Figure 5.18 Pink (a) and grey (b) varieties of Delhi quartzites with Index properties of Delhi quartzite

(Source of data: Negi and Chakraborty, 2016)

5.8 Other natural rocks and stones used in the UNESCO monuments of Delhi and Agra

Besides the afore-mentioned stones used in bulk quantities, diverse natural rocks and minerals (semi-precious gemstones) have also been used in subordinate quantities to embellish the edifices inside the Taj Mahal Complex, Agra Fort Complex, Red Fort Complex and Humayun's Tomb Complex (Fig. 5.19). An exhaustive list of the various stones/semi-precious gemstones used in Taj Mahal is given in Table 5.4. The semi-precious stones used in the inlay work in marble in Taj Mahal were procured from different places (Table 5.4; Nath, 1985). Black slate was commonly used to enhance the contrast between the red sandstone and white marble and for inlaying holy Quaranic verses on the white marble façade on the four sides of the main mausoleum inside the Taj Mahal Complex. It was probably brought from Alwar. The yellow limestone of

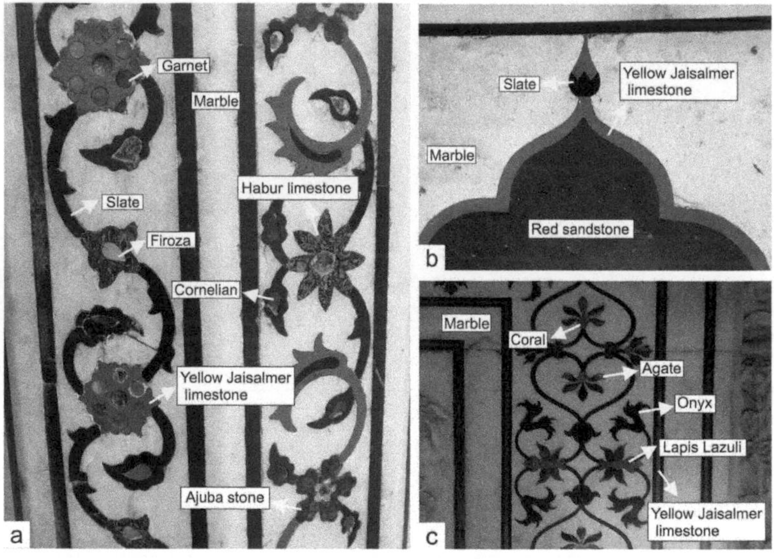

Figure 5.19 Illustrations from parts of (a) Agra Fort Complex and (b and c) Taj Mahal Complex, exhibiting use of various subordinate stones in the UNESCO monuments

(Photos: Gurmeet Kaur and Parminder Kaur)

Table 5.4 List of semi-precious stones and rocks used in the Taj Mahal Complex

S. No	*Name of the stone in Persian*	*Translated in English*	*Places whence received*
1	Aqîq	Cornelian	Baghdad
2	Yeminî	Cornelian	Yemen
3	Firozã	Turquoise	Upper Tibet
4	Lãjward	Lapis-Lazuli	Ceylon
5	Moongã	Coral	Seaside
6	Sulaimãnî	Onyx	South India
7	Ghori		Cambay
8	Lahsûniã	Chrysolite	River Nile
9	Tãmrã	Garnet	River Ganges
10	Tilãî	Golden Stone	Some Mountain
11	Pai-Zahar		Kumaon
12	Sang-i-Gwalior	Stone from Gwalior	Some Mountain
13	Ajûbã	Wonderstone	Surat
14	Sang-i-Moosã	Slate	Jhari
15	Ăbri	Bloodstone	Jhari
16	Khattû	Agate	Jaisalmer
17	Rukham	White marble	Makrana
18	Sang-i-Surkh	Red Stone	Different places
19	Yashab	Jasper	Cambay
20	Pituniã		Cambay
21	Nakhod		Sabalgarh
22	Maknatis	Loadstone	Gwalior

(Source: Nath, 1985)

Jaisalmer and fossiliferous limestone of Habur, used in the inlay work in marble, were procured from localities which belong to present day Rajasthan (Fig. 5.4, 5.5 and 5.6).

Chapter 6

Preservation, conservation and restoration of UNESCO World Heritage Sites of Delhi and Agra

> "*Architecture is the printing press of all ages and gives a history of the state of the society in which it was erected.*"
>
> Lady Morgan (Sydney, 1859). *An Odd Volume Extracted from an Autobiography*, p. 165 (www.azquotes.com/quote/1053261)

6.1 Introduction

In terms of architecture, heritage connotes any building/monument/structure that has a high aesthetic value that not only reflects its peculiar epoch of construction but also transcends that in terms of a timeless appeal. The recognition of an immutable value beyond its immediate historic context, owing to a high aesthetic and historical significance, can be described as architecture with heritage value.

The conservation of built heritage is an activity with far reaching social, environmental and cultural effects. It is a deeply engaging and multi-dimensional field of work that calls for sensitivity towards the existence of heritage in our midst. It involves proper appraisal, a deep professional acumen to aid upkeep and, simultaneously, a commitment to understand materiality of the historical period concerned. In 1810 and 1817, two regulations, namely the "Bengal Regulation" and the "Madras Regulation" were passed by the British administration to vest the government with the power to take over any heritage structure or precincts that required conservation. The "Ancient Monuments Preservation Act" was passed in 1904 which provided the right of state authority in the preservation of the monuments. In 1905, an unprecedented 20 historic structures in Delhi were ordered to be protected. Earlier, with the setting up of the Archeological Survey of India (ASI) in 1861, a mechanism of protection of monuments was put in place. In 1984, with the founding of Indian National Trust for Art and Cultural Heritage (INTACH), the connection with heritage at once acquired popular connotations because of its larger mandate to generate general awareness about heritage (https://cpwd.gov.in/Publication/ConservationHertbuildings.pdf).

Article 2.6 of the INTACH charter makes a differentiation between the Western and Indian approaches to conservation:

> While the Western ideology of conservation advocates minimal intervention, India's indigenous traditions idealise the opposite. Western ideology underpins official and legal conservation practice in India and is appropriate for conserving protected monuments. However, conserving unprotected architectural heritage offers the opportunity to use indigenous practices. This does not imply a hierarchy of either practice or site, but provides a rationale for encouraging indigenous practices and thus keeping them alive. Before undertaking conservation, therefore, it is necessary to identify where one system should be applied and where the other. For this purpose, it is necessary at the outset to make a comprehensive inventory of extant heritage, both tangible and intangible.
>
> (*Source*: http://www.intach.org/about-charter-principles.php, accessed September 23 2019)

6.2 Last 100 years and beyond of conservation activity in India

The conservation and restoration of heritage monuments of India was prevalent long before the formal manuals pertaining to these were written and practiced. During the year 1652, Aurangzeb (son of Emperor Shah Jahan) wrote a letter to Shah Jahan drawing his attention to some problems related to the upkeep of Taj Mahal. The letter primarily points out the crisis of leaking roofs during rainy season in parts of the mausoleum and the adjoining mosque and the assembly hall. In the letter (Firman) Aurangzeb mentions his interaction with the architects on the remedies to preserve the mausoleum by use of mortar. The excerpt given below from Begley and Desai (1989), translated from the original Persian to English, throws light on the conservation strategies followed during those days:

Ruqa'at-I-Alamgiri

This faithful and devoted servant, having respectfully tendered his devotion and obedience, which is the warranty deed of perpetual happiness, humbly brings to [His Majesty's] notice that this disciple (murid) entered Akbarabad on Thursday, the 3rd of the honored month of Muharram 1063 [4 December 1652], that he went directly to the garden of Jahanara, with the intention of calling on that princess of the people of the world. And having enjoyed the bounty of her company in that pleasant abode, toward the close of the day, he returned to the mansion (manzil) situated in the garden of Mahabat Khan. And on Friday, by going to perform the pious circuit (tawaf) of the Illumined Tomb, he obtained the blessings of a visit of complete devotion.

The buildings ('imarat) of this shrine enclosure (hazira) of holy foundation are still firm and strong (ustwar), exactly as they were completed under [His Majesty's]illumined presence, except that the dome over the fragrant sepulcher (marqad) leaks during the rainy season in two places on the north side. Likewise, the four arched portals (pishtaq), several of the recessed alcoves (shahnashinan) of the second story (martaba), the four small domes, the four northern vestibules (suffa) and sub-chambers of the seven-arched plinth (kursi-i-habtdar) have become dampened (darnam). The marble-covered terrace of the large dome has leaked in two or three places during this past rainy season, and has been repaired. Let us see what happens in the coming rainy season. The domes of the mosque and the assembly-hall (jama'at-khana) leaked in the rainy season as well, and have also been repaired.

The architects (bannayan) are of the opinion that if the floor (farsh) of the roof (pusht-i-bam) of the second story

> (martaba) is opened up and treated with concrete grout (rekhta), and above the concrete at half-yard (gaz) layer (tahkari) [of mortar?] is placed, the portal, the gallery and the smaller domes will probably be rectified [i.e., made water-tight]. But they acknowledge their inability [to suggest any remedy] in setting in order the large dome.
>
> (Begley and Desai, 1989)

6.3 Current status of conservation, preservation and restoration practices

Evolving practice of conservation in India thereafter seems to consider a broader range of charters nationally and internationally, namely the ASI's Conservation Manual (Marshall, 1923), UNESCO's operational guidelines, Venice charter, ICOMOS amongst others. Preservation, as the name implies, is aimed at "maintaining a place in its existing state and retarding deteriorations" (Nanda, 2017). To ensure the best practices of conservation and preservation, alterations seen in the last century that seem to be accelerating decay and obscuring original material were consistently and carefully removed and replaced with traditional materials.

The process of conservation, preservation and restoration can be done by first carefully carrying out the documentation and archival research by a trained inter-disciplinary set of professionals. Parts of the alterations carried out in the 20th century in most cases were considered inappropriate in a manner that is not acceptable today. For instance in case of incised plaster, upon due consultation, layers of repaired cement plaster would have to be first removed and later restored back to the closest materiality of its original application. Therefore the process of reconstruction in its original form is required as part of the conservation action plan where through neglect, vandalism or structural failure, significant building have collapsed (Nanda, 2017). Last of all, the conditions of "integrity" and "authenticity" are applied as described in the charters of conservation.

The 20th century saw the introduction of the building material "cement" and its easy availability in the Indian scenario for construction purposes. One of the quick applications was seen in repair of monuments where some intervention was required. At the outset, some downsides of using cement were unavoidable. Usage of cement not only obliterates the historic character of the monument but also is not compatible with lime plaster which has been originally used in almost all historic structures. The cement is also corrosive to some soft sandstone, which did not particularly help.

Also for conservation purposes, rampant use of chemicals for cleaning the surface of stone started in the early part of 20th century but a few decades down the line, the ill effects of such materials became apparent and a rather glaring mistake of its usage over years seems irreversible. As awareness slowly crept in, most of the repair works for waterproofing, plastering, pointing of stone joints, flooring and so on came to be seen as bad practices in the conservation disciplines.

6.4 Current state of conservation, preservation and restoration of UNESCO Heritage Sites of Delhi and Agra

The UNESCO monuments of Delhi and Agra have suffered degradation and deterioration, with the passage of time due to multiple reasons viz., change of dynasties, vandalism, natural process of wear and tear, urbanization, industrial setup in proximity to these monuments etc. (Batra, 2007). The effects of deterioration, on stones used in all the monuments discussed in the preceding chapters of this book, are visible in different segments of these monuments, and their restoration is a complicated task. The stones used in these monuments exhibit damage/deterioration due to vandalism (Fig. 6.1a), natural process of wear and tear commonly termed as weathering (Fig. 6.1b and c), color change (Fig. 6.1c), rock failure (Fig. 6.1d), insensitive public intervention (Fig. 6.2a and c), exfoliation caused by physical/chemical weathering (Fig. 6.2b) and chipping out of semi-precious stones used in inlay work owing to varied reasons (Fig. 6.2d).

The UNESCO monuments are under the ambit of the ASI for maintenance purposes. The restoration of the decaying and deteriorating parts of the stone heritage is a cumbersome and tricky task and requires expertise from varied groups working in this direction, as has been discussed in the preceding section of this chapter. The fair examples of restoration and preservation by ASI can be observed at Humayun's Tomb Complex, Red Fort Complex, Taj Mahal and Agra Fort Complex where the restoration and conservation process has been underway since the early 20th century and is still continuing. Perhaps the most significant example

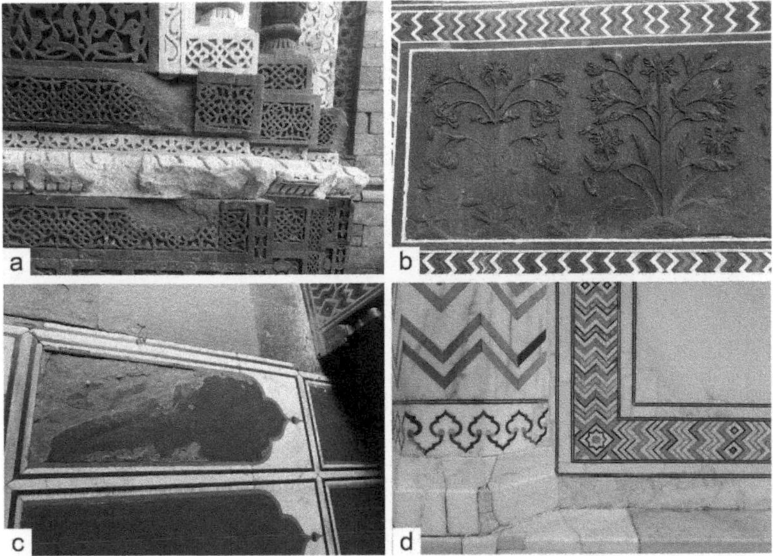

Figure 6.1 (a) Damaged marble and red sandstone carved gateway of Alai Darwaza of Qutb Complex; (b) weathering of red sandstone dado in the south gateway of Taj Mahal Complex; (c) weathered and decolorized red sandstone floor in the mosque in Taj Mahal Complex; (d) cracks developed in the marble used in the façade of the masoleum in Taj Mahal

(Photos: Gurmeet Kaur and Anuvinder Ahuja)

to consider is the Humayun's Tomb Complex. It was in a dilapidated state for many years as it was used as a refugee camp by the displaced people during India-Pakistan Partition in 1947 (www.india-seminar.com/2011/628/628_ratish_nanda.htm). It was a big blow to the complex, and its vital components were badly damaged. The conscience and timely efforts of ASI in concert with Agha Khan Trust and INTACH in the past two decades have been instrumental in reviving this monument much to its original form (Nanda, 2017). In Red Fort Complex, Agra Fort Complex and Taj Mahal Complex one can witness the revival of different segments in a phased manner where restoration of the original components has been achieved to a satisfactory level (Fig. 6.3a–d). The worn out tracery, lattice work and inlay work in stone in many segments of these monuments have been successfully completed (Fig. 6.3a) and

Figure 6.2 (a) Damaged marble screen at Musmman Burj enclosure; (b) weathered red sandstone pillar in the inner court of Jahangir Mahal; (c) damaged red sandstone screen in Humayun's Tomb; (d) vandalized inlay work in Agra Fort

(Photos: Anuvinder Ahuja, Parminder Kaur and Gurmeet Kaur)

in some the restoration is underway by the skilled and seasoned stone workers (Fig. 6.3b–d). Utmost care is being taken in terms of using the stone material for restoration (personal communication) with Conservation Archaeologist (Red Fort Complex); and Ujwala Menon (Conservation Architect, Agha Khan Trust). Although the restorations of these monuments are being dealt with in a phased manner, nevertheless, there

Figure 6.3 (a) Restored inlay work in marble in Diwan-i-Khas of Red Fort, Delhi; (b) restoration in progress of one of the side minarets of Taj Mahal; (c) restoration of marble screen in Khas Mahal in Red Fort, Delhi; (d) restoration work in progress at Red Fort, Delhi

(Photos: Parminder Kaur, Anuvinder Ahuja and Jaspreet Saini)

are segments in all the five UNESCO World Heritage Sites in Delhi and Agra which require restorations to their original design and form. There are some which have been restored in a negligent and unmindful manner, and they require immediate attention of ASI. One such example which requires immediate attention is the façade of Alai Darwaza in Qutb Complex. It was restored using lime mortar instead of the original building material viz., marble and red sandstone, thus offering a very unpleasant view to the otherwise magnificent carved Alai Darwaza (Fig. 6.4).

Besides the great initiative by Ministry of Culture of India, Ministry of Tourism of India and Archaeological Survey of India, there are many

NGOs and local bodies coming together to make people aware of their stone built architectural heritage by organizing heritage walks, initiating talks on heritage monuments, encouraging people to take part in framing strategies related to conservation, preservation and restoration of these monuments.

Figure 6.4 The façade of Alai Darwaza of Qutb Complex offers an unpleasant view on account of its restoration in lime mortar instead of original building materials viz., marble and sandstone

Chapter 7

Conclusions

UNESCO World Heritage Sites from across the world, recognized as valuable monuments, are spread out in all the continents. UNESCO has enlisted certain Cultural, Architectural, Natural and Mixed Sites from different regions across the world, depending on their unique character that showcases an inordinately developed level of art, culture and traditional values. The stone built monuments in particular are a link to the outstanding past, being timeless repositories of design, creative style of architecture and evolution of civilizations. Humans evolved across epochs by waging a duel with nature at every step, for survival. Since earth is made up of a variety of rocks, humans discovered how to use the given medium to their favor by creating architectural heritage that is unmatched for its magnificence even today. These artistic endeavors were made across the globe. Some countries saw a greater surge in architectural development compared to others. The monuments were either rock hewn, monoliths or were built by procuring stone from nearby quarries and sometimes imported from far-flung places to create exquisite pieces of architecture. UNESCO has been shouldering the responsibility since November 1945 and is taking care of the well-being of these treasured heritage antiquities. The UNESCO is mandated towards recognizing this heritage, creating awareness among masses and planning strategies for preserving, conserving and restoring them for longevity.

The IUGS Heritage Stones Subcommission (HSS) took up the initiative of identifying and designating the natural stones as Global Heritage Stone Resource (GHSR) which have played a pivotal role in the making of the architectural heritage around the globe. The important criterion laid down for designation of GHSR incorporates the use of stone in the historical past, wider geographic relevance, regional and global recognition of the architectural heritage with cultural link, present state of active or historical quarries etc. In the current scenario, a total of 22 stones have been designated as GHSRs with many waiting for their turn to be recognized as GHSRs. Until recently, most of the designated GHSRs are from Europe. Makrana marble form India is the only stone to have been designated as a GHSR in spite of India's rich and culturally diverse stone built heritage.

This book is second in the series on Natural Stone and World Heritage where we bring together a comprehensive study of monuments included in the UNESCO World Heritage list from Delhi and Agra. A multi-pronged approach is adopted in this book, focusing on the heritage stones, geological, historical, architectural and aesthetic-cultural aspects of five UNESCO designated World Heritage Sites, namely The Qutb Minar and adjoining monuments, Humayun's Tomb, Red Fort Complex, Agra Fort and Taj Mahal. Not only is each of these monuments a product of high human ingenuity and creativity, they collectively represent a unique historical-cultural idiom in the history of human evolution. Architecturally, aesthetically and politically, these monuments stand out as bearers of a unique point in Indian history where human efforts collated to produce tangible heritage that has outlived centuries.

These stone built monuments help us connect the dots for what was a very complex, yet fertile period of Indian history. It indicates the nascent Islamic influence: the very first mosque in India extant in the Qutb Complex, which in itself is a pastiche of influences, patched together with varied aesthetic sensibilities, executed by artisans unfamiliar in a foreign vocabulary. Gradually, however, a more rooted style began to emerge which amalgamated aesthetics of Persian design and Indian craftsmanship, motifs and geological materials. This confluence can be seen to be the progenitor of valuable stone heritage that emerged out of these civilizational encounters. It led to the creation of a third entity that can be said to belong to the soil. Much like Urdu emerged as a language of the soldiers in army encampments where a motley bunch spoke languages of diverse language families, Urdu functioned like a bridge that provided a *lingua franca*. And subsequently what emerged to ease communication ended up touching a pinnacle of refinement in poetry and ideas.

Similar was the case with architectural stone built heritage that was produced as a result of these encounters. What began as a violent imposition, over time, began to grow roots in the new lands. If this was an indicator of new power brokers, it was equally a promise of novel influences. By and by, the violent control gave way to more and more indigenization. While design templates were exported from the lands of the conquerors, it was not always possible to translate those ideas using the old materiality. So an inventive and robust exchange of labor, craftsmanship and materiality ensued, leading to fecund experimentation and subsequent creation of stone structures, motifs and styles that illustrated the terms of this exchange. The idea, therefore, of using indigenous stones to create novel structures and vice versa is tangible in these monuments. By and by these influences got mixed and the initial foreignness gave way to a more grounded new reality, which could have roots in another land but was throwing shoots and aerial roots too.

These monuments span a time frame of five centuries: with the earliest, Qutb, dating 12th century to Taj Mahal in the 17th century. If one were to look at these solely in an architectural light, they span an aesthetic, the initiation, development and prodigious flowering of which is all visible in the five stone built monuments. From the tentative pastiche-like character of structures in the Qutb Complex to the refinement of conventions like *chaarbagh* and garden tombs, palpable in Humayun's Tomb, to its full development in Taj Mahal, one can see a full circle.

Architecturally, the Qutb saw one of the earliest examples of Islamic architecture in India. With these styles being tried on Indian soil for the first time, rather unmediated beginnings of a fascinating era of Indo-Islamic architecture were made. The conventional material palette of the time broadened to include veneering choices like the red sandstone and white marble with surface decorations and carvings. Hindu elements were fused with the traditional Islamic architectural style to generate an architectural vocabulary specific to the Indian context. The value of the monument lies in its illustrative value as a pastiche, which over time saw internal development owing to a greater engagement and addition of more refined structures to the original. A range of this development is visible within the precincts itself. The Qutb Minar and the adjoining monuments were mostly constructed in Delhi quartzite, varieties of Vindhyan Sandstone, marble and Alwar slate. The Vindhyan Sandstone used in the Qutb Complex exhibit vivid textural and color variations. The hues of beige, red, yellow and banded varieties of sandstones used in these monuments impart a unique character to these structures.

Humayun's Tomb saw the first-of-its-kind garden tomb example in the Indian subcontinent. The *chaarbagh* concept of Islamic gardens was adopted in the model of a dynastic tomb. It is also said that no earlier precedent of this nature was present back in history anywhere in the world – in scale, grandeur, order and symmetry. Humayun's Tomb also saw a greater use of ornamentation with material inlays, colored exterior tile cladding as seen in *Nilu Gumbad*, incised plaster decorations and so on. It also started to show innovation in structural systems deployed – for instance, squinches and rest double domes. This is also the very first structure in India to demonstrate a double dome system which originally came from Persia. The building stone predominantly used in the Humayun's Tomb was red, beige and yellowish Vindhyan Sandstone from Tantpur. The white Makrana marble was mostly used for inlay work in the sandstones and cladding the dome of the tomb, besides being used in the cenotaphs inside the Humayun's Tomb Complex. The greyish Delhi quartzite was part of the plinth of the Tomb structure with panels of red sandstone. Minor use of Alwar slates as an inlay material imparts a distinct character to the façade of the tomb.

The Agra Fort saw an articulated urban planning in a fortress setting, with various functions coming together in a cohesive urban plan. The complex is a layered space, as if illustrating the historical antecedents of the city of Agra itself in miniature. The fort, in its present form, becomes a resting space of the myriad architectural and historical influences that touch the monumental scale of Mughal architecture. Important building stones used in Agra Fort include red and spotted Vindhyan Sandstone varieties from Bansi Paharpur and Rupbas, marble from Makrana, yellow and Habur limestone from Jaisalmer and Habur, and slates from Alwar region. Semi-precious stones used mostly for intricate inlay work in the Khas Mahal, Diwan-i-Khas and Musamman Burj were brought from distant places. Precious stones such as rubies, diamonds, emeralds and sapphires were used in the Peacock throne which was later shifted to the Diwan-i-Khas of Red Fort of Delhi.

Taking on from the Agra Fort, the Red Fort saw further refinements of architectural typologies – the fortification structures, public meeting halls, royal residences and so on. Interestingly, it is said to imbibe the Mughal, Persian, Hindu and Timurid styles in some way or other. The material palette also broadened with a superior level of detail and the craft of ornamentation to include mirror work and white shell plasters for a seamless finish. Red sandstone was extensively used with white marble and inlays of *pietra dura* and other semi-precious and precious stones such as coral, lapis lazuli, pearl, malachite, lodestone, cornelian, rubies, emeralds etc. particularly in special areas like the private halls of the emperor. The use of Jaisalmer yellow limestone, fossiliferous Habur limestone and Alwar slate for inlay work is also commonly observed in the Khas Mahal, Shah Burj, Diwan-i-Khas, Rang Mahal and Diwan-i-Am.

We culminate our volume with the Taj Mahal Complex. Considered the finest example and artistic achievement of the time, the Taj showcases a high point in conceptualizing and realizing a monument of this stature, which is also amongst the most admired world monuments today. It combined the finest workmanship possible in the multiple areas of architecture to horticulture to several crafts, and these multiple interventions were executed with an engineer's acumen and jeweller's finesse. The Taj Mahal is considered the world's most well-preserved and architecturally beautiful tomb. Shah Jahan's genius in discerning art and vision drawn up with his architects created a grand timeless monument. The Makrana white marble from Rajasthan was the vital stone of this mausoleum which took almost 17 years for its completion, lending it a unique vocabulary. The red and spotted sandstone procured from the nearby exposures of the Vindhyan Basin in contrast to the marble from neighboring Aravalli Mountain Belt have been extensively used in other monuments adjoining the grand mausoleum. Besides the significant quantities of marble and sandstone, subordinate quantities of Alwar slates, Jaisalmer yellow limestone and Habur limestone have been used in inlay work along with semi-precious stones such as cornelian, agate, onyx, lapiz lazuli, ajuba stone, coral, malachite, magnetite etc. which were procured from far flung places such as Ceylon, Baghdad, Tibet, River Nile and Yemen.

To sum up, one view can look for points of conflict, but it pays to look at the destruction of old monuments and the erection of new ones as a play in time. Poets and litterateurs have upheld this philosophical orientation across ages and it is rewarding because in all fairness, that is the only way to transmute conflict, which has been an intrinsic part of history. It has been a power play with dynasts constantly uprooting one another, inflicting destruction and erecting symbols of their own power. However, in this play is also seen an establishment of one worldview over another. These monuments echo that archetypal stance of the *Nataraja*, that mythic Shiva dance that embraces creation and destruction as twin powers, not as polar opposites but as complementing entities, one paving the way for the other. In the background is the incessant beat of the drum symbolizing the passage of time that, oblivious to the changing vanguard, marches on with stoic indifference.

References

Balasubramaniam, R. (2005) *The World Heritage Complex of the Qutub*. Aryan Books International, New Delhi.

Banerjee, A. & Banerjee, D.M. (2010) Modal analysis and geochemistry of two sandstones of the Bhander group (Late Neoproterozoic) in parts of the central Indian Vindhyan basin and their bearing on the provenance and tectonics. *Journal of Earth System Science*, 119(6), 825–839.

Batra, N.L. (2007) *Dilli's Red Fort by the Yamuna*. Niyogi Books, New Delhi.

Begley, W.E. & Desai, Z.D.A. (eds) (1989) *TajMahal: The illumined tomb. An Anthology of Seventeenth Century Mughal and European Documentary Sources*, The University of Washington Press, Seattle and London.

Bhadra, B.K., Gupta, A.K., Sharma, J.R. & Choudhary, B.R. (2007) Mining activity and its impact on the environment: study from Makrana marble and Jodhpur sandstone mining areas of Rajasthan. *Journal of Geological Society of India*, 70(4), 448–557.

Bhardwaj, B.D. (1970) *Upper Vindhyan Sedimentation in the Kota-Rawatbhata Area, Rajasthan*. PhD Diss., Aligarh Muslim University.

Blochmann, H. (tr) (1873) *The Ain-I Akbariby Abu'l-Fazl Allami*, Volume I. The Baptist Mission Press, Calcutta.

Blochmann, H. (tr) & Phillot, D.C. (ed.) (1927, reprint 1993) *The Ain-I Akbari by Abu'l-Fazl Allami*, Volume I. Asiatic Society of Bengal, Calcutta.

Bose, P.K. & Chakraborty, P.P. (1994) Marine to fluvial transition: Proterozoic upper Rewa Sandstone, Maihar, India. *Sedimentary Geology*, 89(3–4), 285–302.

Bose, P.K., Sarkar, S., Chakrabarty, S. & Banerjee, S. (2001) Overview of the meso-to neoproterozoic evolution of the Vindhyan Basin, Central India. *Sedimentary Geology*, 141, 395–419.

Byerly, D.W. & Knowles, S.W. (2017) Tennessee "marble": a potential "global heritage stone resource". *Episodes*, 40(4), 325–331.

Careddu, N. & Grillo, S.I.L.V.A.N.A. (2015) Rosa Beta granite (Sardinian Pink Granite): A heritage stone of international significance from Italy. *Geological Society of London, Special Publication*, 407(1), 155–172.

Cassar, J., Torpiano, A., Zammit, T. & Micallef, A. (2017) Proposal for the nomination of lower globigerina limestone of the Maltese Islands as a "global heritage stone resource". *Episodes*, 40(3), 221–231.

Cavallo, G. & Pandit, M. (2008) Geology and petrography of ochres and white clay deposits in Rajasthan state (India). *Proceedings of the Conference Geoarchaeology and Archaeomineralogy*, 52, 147–152.

Chakraborti, P.P., Dey, S. & Mohanty, S. (2010) Proterozoic platform sequences of peninsular India: implications towards basin evolution and supercontinent assembly. *Journal of Asian Earth Science*, 39, 589–607.

Chopra, P. (ed.) (1976) *Delhi Gazetteer*. The Gazetteer Unit, Delhi Administration, New Delhi.

Cole, H.H. (1873) *The Architecture of Ancient Delhi*. Arundel Society, London.

Cravero, F., Ponce, M.B., Gozalvez, M.R. & Marfil, S.A. (2015) Piedra Mar del Plata': an argentine orthoquartzite worthy of being considered as a global heritage stone resource. *Geological Society of London, Special Publication*, 407(1), 263–268.

Deb, M., Thorpe, R.I., Krstic, D., Corfu, F. & Davis, D.W. (2001) Zircon U-Pb and galena Pb isotope evidence for an approximate 1.0 Ga terrane constituting the Western margin of the Aravalli-Delhi orogenic belt, Northwestern India. *Precambrian Research*, 108, 195–213.

De Kock, T., Boone, M., Dewanckele, J., DeCeukelaire, M. & Cnudde, V.(2015) Lede stone: a potential "global heritage stone resource" from Belgium. *Episodes*, 38(2), 91–96.

Dube, R.K. (2008) Superiority of Makrana (Rajasthan) marble. *Indian Journal of History of Science*, 43(2), 285–288.

DuTemple, L.A. (2003) *The TajMahal*. Twenty-First Century Books, Minneapolis.

Erskine, K.D. (1908) *Imperial Gazetteer of India Provincial Series: Rajputana*. Superintendent of Government Printing, Calcutta.

Folk, R.L. (1980) *Petrology of Sedimentary Rocks*. Hemphill Publishing Company, Austin.

Foster, W. (ed.) (1921) *Early Travels in India, 1583–1619*. Oxford University Press, London.

Fratini, F., Pecchioni, E., Cantisani, E., Rescic, S. & Vettori, S. (2015) Pietra serena: the stone of the renaissance. *Geological Society of London, Special Publication*, 407(1), 173–186.

Freire-Lista, D.M., Fort, R. & Varas-Muriel, M.J. (2015) Alpedrete granite (Spain). A nomination for the "global heritage stone resource" designation. *Episodes*, 38(2), 106–113.

Frowde, H. (1908) *The Imperial Gazetteer of India*, Oxford Clarendon Press, London.

Garcia Talegon, J., Iñigo, A.C., Alonso-Gavilán, G. & Vicente-Tavera, S. (2015) Villamayor stone (golden stone) as a global heritage stone resource from Salamanca (NW of Spain). *Geological Society of London, Special Publication*, 407(1), 109–120.

Garg, S., Kaur, P., Pandit, M., Kaur, G., Kamboj, A. & Thakur, S.N. (2019) Makrana marble: a popular heritage stone resource from NW India. *Geoheritage*, 11(3), 909–925.

Gilleaudeau, G.J., Sahoo, S.K., Kah, L.C., Henderson, M.A. & Kaufman, A.J. (2018) Proterozoic carbonates of the Vindhyan Basin, India: chemostratigraphy and diagenesis. *Gondwana Research*, 57, 10–25.

Gregory, L.C., Meert, J.G., Pradhan, V., Pandit, M.K., Tamrat, E. & Malone, S.J. (2006) A paleomagnetic and geochronologic study of the Majhgawan Kimberlite, India: implications for the age of the upper Vindhyan supergroup. *Precambrian Research*, 149, 65–75.

Guha, J. & Roonwal, G. (2014) *Stones: Silent Witness to the Cultural Diversity of India*. Kaveri Book Service, New Delhi.

Gupta, B.C. (1934) *The Geology of Central Mewar*. Office of the Geological Survey of India, Calcutta.

Gupta, S.N. (1981) *Delhi Between Two Empires*. Oxford University Press, New Delhi.

Gupta, S.N., Arora, Y.K., Mathur, R.K., Iqbaluddin, P.B., Sahai, T.N. & Sharma, S.B. (1981) *Lithostratigraphy Map of Aravalli Region, Southern Rajasthan and North Eastern Gujarat*. Geological Survey of India Publication, Hyderabad.

Gupta, S.N., Arora, Y.K., Mathur, R.K., Iqbaluddin, P.B., Sahai, T.N. & Sharma, S.B. (1997) Precambrian geology of the Aravalli region, Southern Rajasthan and Northeastern Gujarat, India. *Memoirs of Geological Society of India*, 123, 262.

Habib, I. (1982) *An Atlas of the Mughal Empire: Political and Economic Maps with Detailed Notes, Bibliography and Index*. Centre of Advanced Study in History, Aligarh Muslim University, Aligarh, New Delhi; Oxford University Press, New York.

Hearn, G.R. (1906) *The Seven Cities of Delhi*. W. Thacker & Co., London.

Heldal, T., Meyer, G.B. & Dahl, R. (2015) Global stone heritage: Larvikite, Norway. *Geological Society of London, Special Publication*, 407(1), 21–34.

Herbert, E. (2012) Curzon Nostalgia: landscaping historical monuments in India. *Studies in the History of Gardens & Designed Landscapes*, 32(4), 277–296.

Heron, A.M. (1917a) The Biana-Lalsot hills in Eastern Rajputana. *Records of Geological Survey of India*, 48(4), 181–203.

Heron, A.M. (1917b) Geology of Northeastern Rajputana and adjacent districts. *Memoirs of Geological Survey of India*, 45(1), 1–128.

Heron, A.M. (1953) The geology of central Rajputana. *Memoirs of Geological Survey of India*, 79.

Hughes, T., Horak, J., Lott, G. & Roberts, D. (2016) Cambrian age Welsh slate: a global heritage stone resource from the United Kingdom. *Episodes*, 39(1), 45–51.

Hughes, T., Lott, G.K., Poultney, M.J. & Cooper, B.J. (2013) Portland stone: a nomination for "global heritage stone resource" from the United Kingdom. *Episodes*, 36(3), 221–226.

Hussain, M.A. (1937) *An Historical Guide to the Agra Fort*. Manager of Publications, New Delhi.

Kaur, G., Makki, M.F., Avasia, R.K., Bhusari, B., Duraiswami, R.A., Pandit, M.K., Baskar, R. & Kad, S. (2019a) The late cretaceous-paleogene deccan traps: a potential global heritage stone province from India. *Geoheritage*, Springer Berlin Heidelberg, 11(3), 973–989.

Kaur, G., Singh, S., Kaur, P., Garg, S., Pandit, M.K., Agrawal, P., Acharya, K. & Ahuja, A. (2019b) Vindhyan sandstone: a crowning glory of architectonic heritage from India. *Geoheritage*, Springer Berlin Heidelberg, 1–13. https://doi.org/10.1007/s12371-019-00389-8

Khan, A.A. (2013) Paleogeography of the Indian Peninsula vis-à-vis geodynamic and petrotectonic significance of the Vindhyan Basin with special reference to neo-mesoproterozoic. *Journal of Indian Geological Congress*, 5(1), 65–76.

Kramar, S., Bedjanič, M., Mirtič, B., Mladenović, A., Rožič, B., Skaberne, D., Gutman, M., Zupančič, N. and Cooper, B. (2015) Podpeč limestone: a heritage stone from Slovenia. *Geological Society of London, Special Publication*, 407(1), 219–231.

Lopes, L. & Martins, R. (2015) Global heritage stone: estremoz marbles, Portugal. *Geological Society of London, Special Publication*, 407(1), 57–74.

Majid, A., Ahmad, A.H.M. & Bhat, G.M. (2012) Facies controlled porosity evolution of the neoproterozoic Upper Bhander sandstone of Western India. *Geological Society of London, Special Publication*, 366(1), 91–110.

Malone, S.J., Meert, J.G., Banerjee, D.M., Pandit, M.K., Tamrat, E., Kamenov, G.D., Pradhan, V.R. & Sohl, L.E. (2008) Paleomagnetism and detrital zircon geochronology of the upper Vindhyan sequence, Son Valley and Rajasthan, India: a ca. 1000 ma closure age for the Purana Basins? *Precambrian Research*, 164(3–4), 137–159.

Marker, B.R. (2015) Bath stone and purbeck stone: a comparison in terms of criteria for global heritage stone resource designation. *Episodes*, 38(2),118–123.

Marshall, J. (1923) *Conservation Manual.* Superintendent Government Printing, Calcutta.

McKenzie, N.R., Hughes, N.C., Myrow, P.M., Banerjee, D.M., Deb, M. & Planavsky, N.J. (2013) New age constraints for the proterozoic Aravalli-Delhi successions of India and their implications. *Precambrian Research*, 238, 120–128.

Meert, J.G., Pandit, M.K., Venkateshwarlu, M. & Rao, J.M. (2013) Comment: paleomagnetism of Bhander sediments from Bhopal Inlier, Vindhyan supergroup. *Journal of Geological Society of India*, 82, 588–589.

Misra, A.K. & Mishra, A. (2007) Escalation of salinity levels in the quaternary aquifers of the Ganga Alluvial Plain, India. *Environmental Geology*, 53(1), 47–56.

Mitra, S. (ed.) (2002) *Qutb Minar and Adjoining Monuments*. Archeological Survey of India, New Delhi.

Nanda, R. (2017) *Humayun's Tomb Conservation.* Agha Khan Trust for Culture & Mapin Publishing, New Delhi.

Naqvi, S.A.A. (2002) *Humayun's Tomb and Adjacent Monuments*. Archeological Survey of India, New Delhi.

Natani, J.V. (2000) Geoenvironmental impact assessment studies of Makrana marble mining area, Nagaur district, Rajasthan. *Journal of Geological Society of India*, 133(7), 64–65.

Natani, J.V. (2002) Regional assessment of marble and calc-silicate rocks of Rajasthan. *Journal of Geological Society of India*, 135(7), 3–55.

Natani, J.V. & Raghav, K.S. (2003) Environmental impact of marble mining around Makrana, Nagaur district, Rajasthan. *Journal of Geological Society of India*, 62(3), 369–376.

Nath, R. (1972) *The Immortal Taj Mahal: The Evolution of the Tomb in Mughal Architecture*. D. B. Taraporevala Sons, Bombay.

Nath, R. (1985) *The Taj Mmahal and Its Incarnation*. Historical Research Documentation Programme, Jaipur.

Navarro, R., Pereira, D., Cruz, A.S. & Carrillo, G. (2019) The significance of "white macael" marble since ancient times: characteristics of a candidate as global heritage stone resource. *Geoheritage*, 11(1), 113–123.

Negi, P. & Chakraborty, T. (2016) Acoustic emission monitoring on Delhi quartzite under compressive loading. In *Recent Advances in Rock Engineering (RARE 2016)*. Atlantis Press, Paris.

Page, J.A. (1926) *Memoirs of the Archaeological Society of India: An Historical Memoir on the Qutb. Delhi*. Government of India Central Publication Branch, Calcutta.

Page, J.A. & Sharma, Y.D. (2002) *Qutub Minar and Adjoining Monuments*. The Director General, Archaeological Survey of India, New Delhi.

Paliwal, B.S., Pareek, U.S. & Vyas, A. (1977) Structural frame work of the Makrana marble deposit and its bearing on the Precambrian stratigraphy of Rajasthan. *Geology in South Asia-II: Geological Survey and Mines Bureau, Sri Lanka, Professional Paper*, 7, 123–136.

Peck, L. (2005) *Delhi, A Thousand Years of Building*. Roli Books Pvt. Ltd., New Delhi.

Pereira, D. (2019) *Natural Stone and World Heritage: Salamanca (Spain)*. CRC Press, Boca Raton and London.

Pereira, D., Tourneur, F., Bernáldez, L. & Blazquez, G. (2015a) Petit granit: a Belgian limestone used in heritage, construction and sculpture. *Episodes*, 38(2), 85–90.

Pereira, D., Marker, B.R., Kramar, S., Cooper, B.J. & Schouenborg, B.E. (eds) (2015b) *Global Heritage Stone: Towards International Recognition of Building and Ornamental Stones*. Geological Society of London, London.

Prasad, B. (1984) Geology, sedimentation and palaeogeography of the Vindhyan supergroup, Southeastern Rajasthan. *Memoirs of Geological Society of India*, 116, 1–2.

Primavori, P. (2015) Carrara marble: a nomination for "global heritage stone resource" from Italy. *Geological Society of London, Special Publication*, 407(1), 137–154.

Quasim, M.A., Ahmad, A.H.M. & Ghosh, S.K. (2017) Depositional environment and tectono-provenance of Upper Kaimur groups and stones, Son Valley, Central India. *Arabian Journal of Geosciences*, 10(1), 4.

Rasmussen, B., Bose, P.K., Sarkar, S., Banerjee, S., Fletcher, I.R. & McNaughton, N.J. (2002) 1.6 Ga U-Pb zircon age for the Chorhat sandstone, lower Vindhyan, India: possible implications for early evolution of animals. *Geology*, 30(2), 103–106.

Ray, J.S., Veizer, J. & Davis, W.J. (2003) C, O, Sr and Pb isotope systematics of carbonate sequences of the Vindhyan supergroup, India: age, diagenesis, correlations and implications for global events. *Precambrian Research*, 121(1–2), 103–140.

Ray, J.S., Martin, M.W., Veizer, J. & Bowring, S.A. (2002) U-Pb zircon dating and Sr isotope systematics of the Vindhyan supergroup, India. *Geology*, 30(2), 131–134.

Rezzavi, S.A.N. (2010) The mighty defensive fort: Red Fort at Delhi under Shahjahan- its plan and structures as described by Muhammad Waris. *Proceedings of the Indian History Congress*, 71, 1108–1121.

Richardson, L. (1992) *A New Topographical Dictionary of Ancient Rome*. JHU Press, Baltimore and London.

Rose, W.I., Vye, E.C., Stein, C.A., Malone, D.H., Craddock, J.P. & Stein, S.A. (2017) Jacobsville sandstone: a candidate for nomination for "global heritage stone resource" from Michigan, USA. *Episodes*, 40(3), 213.

Roy, A.B. (2006) Proterozoic lithostratigraphy and geochronologic framework of the Aravalli mountains and adjoining areas in Rajasthan and neighbouring states, Northwest India. *Indian Journal of Geology*, 78(1–4), 5–18.

Roy, A.B. & Jakhar, A.R. (2002) *Geology of Rajasthan (Northwestern India) Precambrian to Recent*. Scientific Publishers, Jodhpur.

Roy, A.B. & Kataria, P. (1999). Precambrian geology of the Aravalli mountains and neighbourhood: analytical update of recent studies. In: Kataria, P. (ed) *Proceedings of the Seminar on Geology of Rajasthan – Status and Perspective*. MLS University, Udaipur,1999. pp. 1–56.

Roy, A.B. & Kröner, A. (1996) Single zircon evaporation ages constraining the growth of the Archaean Aravalli Craton, Northwestern Indian shield. *Geological Magazine*, 133(3), 333–342.

Roy, A.B. & Paliwal, B.S. (1981) Evolution of lower proterozoic epicontinental deposits: stromatolite bearing Aravalli rocks of Udaipur, Rajasthan, India. *Precambrian Research*, 14(1), 49–74.

Roy, K. (1966) *One Hundred and One Poems by Rabindranath Tagore*. Asia Book House, Bombay.

Saha, D. & Mazumdar, R. (2012). An overview of the palaeoproterozoic geology of Peninsular India, and key stratigraphic and tectonic issues. *Geological Society of London, Special Publication*, 365, 5–29.

Sanderson, G. & Shuaib, M. (2000) *Delhi Fort: A Guide of the Buildings and Gardens*. Asia Educational Services, New Delhi.

Sarkar, S., Eriksson, P.G. & Chakraborty, S. (2004) Epeiric sea formation on neoproterozoic supercontinent break-up: a distinctive signature in coastal storm bed amalgamation. *Gondwana Research*, 7(2), 313–322.

Schouenborg, B., Andersson, J., Göransson, M. & Lundqvist, I. (2015) The Hallandia gneiss, a Swedish heritage stone resource. *Geological Society of London, Special Publication*, 407(1), 35–48.

Sen, S., Mishra, M. & Patranabis-Deb, S. (2014) Petrological study of the Kaimur group sediments, Vindhyan supergroup, Central India: implications for provenance and tectonics. *Geosciences Journal*, 18(3), 307–324.

Sharma, N.L. (1953) *Problems in the correlation of the pre-Vindhyan igneous rocks of Rajasthan*. Presidential Address. 40th Indian Science Congress, Lucknow, 1–28.

Sharma, Y.D. (2001) *Delhi and Its Neighbourhood*. Archeological Survey of India, New Delhi.

Sharma, Y.D. (2015) *Delhi and its Neighbourhood*. Archeological Survey of India, New Delhi.

Siddiqi, W.H. (2008) *Agra Fort: World Heritage Series*. Archeological Survey of India, New Delhi.

Siddiqi, W.H. (2009) *Taj Mahal: World Heritage Series*. Archeological Survey of India, New Delhi.

Silva, Z.C. (2019) Lioz – a royal stone in Portugal and a monumental stone in colonial Brazil. *Geoheritage*, 11(1), 165–175.

Singh, S.P. (1984a) Fluvial sedimentation of the proterozoic Alwar group in the Lalgarh graben, Northwestern India. *Sedimentary Geology*, 39(1–2), 95–119.

Singh, S.P. (1984b) Palaeocurrent and clastic dispersal pattern of the proterozoic Alwar group around Jaipur, Northeastern Rajasthan. *Journal of Geological Society of India*, 25(9), 585–597.

Singh, S.P. (1988) Stratigraphy and sedimentation pattern in the proterozoic Delhi supergroup, Northwestern India. *Memoirs of Geological Society of India*, 7, 193–206.

Sinha-Roy, S. (1984) Precambrian crustal interaction in Rajasthan, NW India. *Proceeding of Seminar on Crustal Evolution of Indian Shield and Its Bearing on Metallogeny*, Indian Journal of Earth Sciences, pp. 84–91.

Sinha-Roy, S., Malhotra, G. & Mohanty, M. (1998) *Geology of Rajasthan*. Geological Society of India, Bangalore.

Smith, V.A. (1999) *The Early History of India*. Atlantic Publishers and Distributors, New Delhi.

Spear, P. (1949) Bentink and the Taj. *The Journal of the Royal Asiatic Society of Great Britain and Ireland*, 2, 180–187.

Spear, T.G.P. (1997) *Delhi: Its Monuments and History*, 3rd Edition. Updated and Annotated by Narayani Gupta & Laura Sykes. Oxford University Press, London.

Spear, T.G.P. (1943) *Delhi: Its Monuments and History*. Oxford University Press, London.

Srivastava, R.K. (1988) Magmatism in the Aravalli mountain range and its environs. *Memoirs of Geological Society of India*, 7, 77–94.

Temple, L.C.S.R.C. (ed.) (1914) *The Travels of Peter Mundy, in Europe and Asia, 1608–1667: Travels in Asia, 1628–1634*, Volume 2, Second Series, no. xxxv. Hakluyt Society, London.

Thussu, J.L. (2006) *Geology of Haryana and Delhi*. Geological Society of India Publication, Bangalore.

Tripathi, J.K. & Rajamani, V. (2003) Geochemistry of proterozoic Delhi quartzites: implications for the provenance and source area weathering. *Journal of Geological Society of India*, 62, 215–226.

Verma, A. (1985) *Forts of India*. Publication Division, New Delhi.

Verma, A. & Shukla, U.K. (2015) Deposition of the upper Rewa sandstone formation of proterozoic Rewa group of the Vindhyan Basin, MP, India: a reappraisal. *Journal of Geological Society of India*, 86(4), 421–437.

Verma, R.K. (1991) *Geodynamics of the Indian Peninsula and the Indian Plate Margin*. Oxford & IBH Publishers Pvt. Ltd., New Delhi.

Wedekind, W., Gross, C.J. & López-Doncel, R. (2017) Rock characteristics and weathering of rock-cut monuments in Lycia (Turkey). *International Symposium on the Conservation of Monuments in the Mediterranean Basin*. Springer, Cham, 2017, pp. 507–514.

Wiedenbeck, M., Goswami, J.N. & Roy, A.B. (1996) Stabilization of the Aravalli Craton at 2.5 Ga: an ion-microprobe zircon study. *Chemical Geology*, 129 (3–4), 325–334.

Wikström, A. & Pereira, D. (2015) The Kolmården serpentine marble in Sweden: a stone found both in castles and people's homes. *Geological Society of London, Special Publication*, 407(1), 49–56.

Online resources

Archeological Society of India. (2018) *Glimpses of Agra Monuments*. Available from: https://ia802807.us.archive.org/7/items/GLIMPSESOFAGRAMONUMENTS/GLIMPSES%20OF%20AGRA%20MONUMENTS.pdf [accessed September 30 2019].

Asher, C.B. (1992) *Architecture of Mughal India*. Cambridge University Press, Cambridge. Available from: https://archive.org/stream/iB_in/1-4_djvu.txt [accessed October 3 2019].

Augustyn, A., Bauer, P. & Duignan, B. *et al.* (2013) *Myra-Turkey*. Available from: www.britannica.com/place/Myra [accessed October 3 2019].

Augustyn, A., Bauer, P. & Duignan, B. *et al.* (2017) *Temple Complex, Angkor Wat, Cambodia*. Available from: www.britannica.com/topic/Angkor-Wat [accessed October 3 2019].

Augustyn, A., Bauer, P. & Duignan, B. *et al.* (2018) *Cappadocia Ancient District, Turkey*. Available from: www.britannica.com/place/Cappadocia [accessed October 3 2019].

Augustyn, A., Bauer, P. & Duignan, B. *et al.* (2018) Available from: www.britannica.com/place/Ajanta-Caves. [accessed October 3 2019].

Augustyn, A., Bauer, P. & Duignan, B. *et al.* (2018) Available from: www.britannica.com/place/Elephanta-Island [accessed October 3 2019].

Augustyn, A., Bauer, P. & Duignan, B. *et al.* (2018) Available from: www.britannica.com/place/Ellora-Caves [accessed October 3 2019].

Augustyn, A., Bauer, P. & Duignan, B. *et al.* (2018) *Chittaurgarh, India*. Available from: www.britannica.com/place/Chittaurgarh [accessed October 5 2019].

Augustyn, A., Bauer, P. & Duignan, B. *et al.* (2019) Available from: www.britannica.com/topic/Westminster-Abbey [accessed October 4 2019].

Augustyn, A., Bauer, P. & Duignan, B. *et al.* (2019) Available from: www.britannica.com/place/Vijayanagar [accessed October 5 2019].

Augustyn, A., Bauer, P. & Duignan, B. *et al.* (2019) Available from: www.britannica.com/place/Petra-ancient-city-Jordan [accessed October 3 2019].

Augustyn, A., Bauer, P. & Duignan, B. *et al.* (2019) Available from: www.britannica.com/topic/Sasanian-dynasty [accessed October 3 2019].

Banerjee, A. (2018) *Aravallis: A Neglected Geological Marvel.* Available from: www.livemint.com/Leisure/f2XeUnRMnc1mIlpVZDrZ0I/Aravallis-A-geological-marvel.html [accessed September 30 2019].

Centre for Cultural Resources and Training. (2017) *SherMandal.* Available from: https://nroer.gov.in/55ab34ff81fccb4f1d806025/file/57d9439f16b51c0da00db599 [accessed September 25 2019].

Chittorgarh.com Team. (2018) History of Chittorgarh. Available from: chittorgarh.com/article/chittorgarh-history/231/ [accessed October 5 2019].

Edwards, C. (2019) *Ancient Turkish Rock Carvings that have Baffled Scientists Could be a Calendar.* Available from: https://nypost.com/2019/06/26/ancient-turkish-rock-carvings-that-have-baffled-scientists-could-be-a-calendar/ [accessed September 30 2019].

Fergusson, J. (1864) *The Rock-Cut Temples of India.* Available from: https://archive.org/stream/rockcuttemplesof00ferg#page/n17/mode/2up [accessed October 3 2019].

Guterbock, H.G. (2014) *Bogazkoy, Turkey.* Available from: www.britannica.com/place/Bogazkoy [accessed October 3 2019].

Handwerk, B. *Pyramids at Giza.* Available from: www.nationalgeographic.com/archaeology-and-history/archaeology/giza-pyramids/ [accessed October 3 2019].

Havell, E.B. (1904) *A Handbook to Agra and the Taj, Sikandra, Fatehpur Sikri and the Neighbourhood.* Available from: https://archive.org/details/cu31924024120200/page/n12/mode/2up [accessed October 4 2019].

History.com Editors (2018). Available from: www.history.com/topics/landmarks/angkor-wat [accessed October 3 2019].

Huke, R.E. & Das, M.N. (2019) *Odisha.* Available from: www.britannica.com/place/Odisha/History [accessed October 5 2019].

Jain, S. (2018) *The Cities of Delhi: From the Legend of Indraprastha to Qila Rai Pithora.* Available from: www.hindustantimes.com/delhi-news/the-cities-of-delhi-from-the-legend-of-indraprastha-to-qila-rai-pithora/story-B9mKCh192j5aVEcBUzJnYI.html [accessed September 30 2019].

Jarry, M. (2019) *Periods and Centres of Activity, Ancient Western World.* Available from: www.britannica.com/place/Mogao-Caves [accessed October 3 2019].

Koch, E. (1991) *Mughal Architecture.* Available from: https://archive.org/details/in.gov.ignca.79132/page/n115/mode/2up [accessed October 1 2019].

Latif, S.M. (1896; republished 2003) *Agra: Historical and Descriptive with an Account of Akbar and His Court and of the Modern City of Agra*. Calcutta Central Press Company Ltd, Calcutta. Available from: https://archive.org/details/agrahistoricalde00syad/page/n19/mode/2up [accessed October 25 2019].

Nanda, R. *A Millennium of Building, 50 Years of Destruction*. Available from: www.india-seminar.com/2011/628/628_ratish_nanda.htm [accessed September 30 2019].

Rao, V., Ram, V. & Sundaram, K.V. *Delhi-History*. Available from: www.britannica.com/place/Delhi/History-ref293596 [accessed September 30 2019].

Shaikh, N. (2015) *Repair techniques for conservation of historic structures*. Thesis. https://www.academia.edu/38221450/Repair_techniques_for_conservation_of_historic_structures [accessed September 30 2019].

Sharma, A. (2019) *Marble Used for Taj Mahal is Now "Global Heritage Stone Resource"*. Available from: www.hindustantimes.com/ritan/marble-used-for-taj-mahal-is-now-global-heritage-stone-resource/story-t67WWKE5kj05JL9o3qEdlO.html [accessed October 3 2019].

Sharp, R.N. (2019) *Persepolis*. Available from: www.britannica.com/place/Persepolis#ref31169 [accessed October 3 2019].

Singh, B.P. (1996) *Indian Archaeology 1991–1992 – A Review*. Archaeological Survey of India, Government of India, New Delhi. Available from: http://nmma.nic.in/nmma/nmma_doc/Indian%20Archaeology%20Review/Indian%20Archaeology%201991-92%20A%20Review.pdf [accessed September 25 2019].

Smith, R.V. (2010) *Delhi's Mahabharata Connection*. Available from: www.thehindu.com/news/cities/Delhi/Delhis-Mahabharata-connection/article16837830.ece [accessed September 25 2019].

Steven, M. (2019) *The City with Nine Lives*. Available from: www.telegraph.co.uk/travel/destinations/asia/india/delhi/articles/delhi-nine-lives/ [accessed September 25 2019].

Tankha, M. (2016) *Mahabharat Sites Continue to Have the Same Names Even Today: B. B. Lal*. Available from: www.thehindu.com/news/cities/Delhi/ritannica-sites-continue-to-have-the-same-names-even-today-b-b-lal/article5776270.ece [accessed September 30 2019].

Tankha, M. (2016) *The Discovery of Indraprastha*. Available from: www.thehindu.com/news/cities/Delhi/the-discovery-of-indraprastha/article5772895.ece [accessed September 30 2019].

Tuli, A.C. (2016) *For Delhi, the Heart of India*. Available from: www.thestatesman.com/features/for-delhi-the-heart-of-india-1480115730.html [accessed September 30 2019].

Vijetha, S.N. (2016) *JNU's Rocking Pre-Historic Legacy*. Available from: www.thehindu.com/news/cities/Delhi/JNUs-rocking-pre-historic-legacy/article13377963.ece [accessed September 30 2019].

Wang, M.C. (2018) *Dunhaung Art*. Available from: https://oxfordre.com/religion/abstract/10.1093/acrefore/9780199340378.001.0001/acrefore-9780199340378-e-173?rskey=OkuAqh&result=16 [accessed October 3 2019].

Wolpert, S.A. & Allchin, F.R. (2018) Available from: www.britannica.com/place/India/History-ref484915 [accessed October 3 2019].

Direct weblinks

https://earthexplorer.usgs.gov/; [accessed September 30 2019].

http://adsabs.harvard.edu/abs/2017EGUGA..19.2639P; [accessed September 30 2019].

http://globalheritagestone.com/; [accessed September 30 2019].

http://globalheritagestone.com/igcp-637/igcp-achievements/; [accessed September 30 2019].

http://globalheritagestone.com/other-projects/ghsr/designations/; [accessed September 30 2019].

http://globalheritagestone.com/reports-and-documents/terms-of-reference/; [accessed September 30 2019].

http://whc.unesco.org/en/criteria/; [accessed September 30 2019].

www.intachdelhichapter.org/listings.php; [accessed September 30 2019].

https://archive.org/; [accessed September 30 2019].

https://archive.org/details/in.ernet.dli.2015.207013/page/n233; [accessed September 30 2019].

https://archive.org/details/in.ernet.dli.2015.207013/page/n245; [accessed September 30 2019].

https://archive.org/details/in.ernet.dli.2015.207013/page/n363; [accessed September 30 2019].

https://archive.org/details/in.ernet.dli.2015.207013/page/n373; [accessed September 30 2019].

https://archive.org/details/in.ernet.dli.2015.207013/page/n379; [accessed September 30 2019].

https://archive.org/details/in.ernet.dli.2015.207013/page/n391; [accessed September 30 2019].

https://archive.org/details/in.ernet.dli.2015.207013/page/n87; [accessed September 30 2019].

https://archive.org/details/in.ernet.dli.2015.279471/page/n279; [accessed September 28 2019].

https://archive.org/details/in.ernet.dli.2015.279471/page/n280; [accessed September 28 2019].

https://archive.org/details/in.gov.ignca.3220/page/n187; [accessed September 28 2019].

https://archive.org/details/travelsofpetermu02mund/page/240; [accessed September 28 2019].

https://archive.org/details/travelsofpetermu02mund/page/n343; [accessed September 28 2019].

https://cpwd.gov.in/Publication/ConservationHertbuildings.pdf [accessed September 15 2019]

https://dsal.uchicago.edu/reference/gaz_atlas_1931/pager.php?object=28; [accessed September 30 2019].

https://ducic.ac.in/cdn/ducic/NewsEventsCommons/GatesofDelhi.pdf; [accessed September 30 2019].

https://nypost.com/2019/06/26/ancient-turkish-rock-carvings-that-have-baffled-scientists-could-be-a-calendar/; [accessed September 30 2019].

https://whc.unesco.org/en/convention/; [accessed September 30 2019].

https://whc.unesco.org/en/list/; [accessed September 30 2019].

https://whc.unesco.org/en/list/231; [accessed September 30 2019].

https://whc.unesco.org/en/list/232; [accessed September 30 2019].

https://whc.unesco.org/en/list/233; [accessed September 30 2019].

https://whc.unesco.org/en/list/251; [accessed September 30 2019].

https://whc.unesco.org/en/list/252; [accessed September 30 2019].

https://whc.unesco.org/en/statesparties/in; [accessed September 30 2019].

www.iugs.org/; [accessed September 30 2019].

www.rkmarblesindia.com/indian-marbles/makrana-marbles; [accessed November 30 2017].

(http://www.unesco.org/new/en/%20unesco/about-us/who-we-are/history/; [accessed September 30 2019].

www.mapsofindia.com; [accessed on 26 September 2019].

Glossary

Amalaka Segmented/ribbed bulbous section resembling an Indian fruit "amala" crowning the top part of shikhara

Archean 4000 to 2500 million years ago

Ayat Ayat is a "verse" of varying length that makes up the chapters of the Quran

Baoli Stepwells are wells or ponds in which the water is reached by descending a set of steps to the water level

Bracket Support element for a projection

Burj Tower

Cenotaph A tomb-like monument to someone buried elsewhere, especially one commemorating people who died in a war

Chaarbagh Quadrilateral garden typically seen in Islamic/Indo-Islamic architecture

Chajjas A chajja is the projecting or overhanging eaves or cover of a roof, usually supported on large carved brackets

Chamfer To cut off the edge/corner to make a symmetrical sloping edge.

Chhaja eaves Projecting or overhanging eaves or cover of a roof, usually supported on large carved brackets

Chhatri Elevated, dome shaped pavilions acting as turrets over roofs

Chilla A spiritual practice in Sufism marked by solitude and penance. Here, the site where the great master Nizamud din Auliya practiced it and later acquired the status of a pilgrimage.

Chunam A type of plaster made with shell lime and sand

Cladding A covering or coating on a structure or material

Corbelled Courses of stone, bricks etc. shifting outward from the main surface which acts as a structural system

Cornice A molded part of the roofline as it touches the ceiling

Cupola A dome-like structure forming/adorning a roof

Dado Lower part of the wall usually finished in a different manner than the rest of the wall

Darwaza Gateway

Dhwaja Stambh Victory standard

Finial A terminating piece at the top, end or corner of an object or at an apex of a roof/dome

Garuda Mythological bird, vehicle of Vishnu

Ghat Quay

Gumbad Dome

Hauz Bath tub

Inlays Embed (something) into the surface of an object

Iwan A rectangular space walled on three sides and usually open on one side and with a vaulted roof

Jali/Jaali Perforated screens, made with carving stone or other materials

Jannah Paradise

Jharokha Overhanging, enclosed balcony

Lapis Lazuli A deep blue metamorphic rock used as a semi-precious stone

Minar Tower/turret

Nahr Canal of water

Naksh and Kufic Styles of Arabic-Islamic calligraphy

Pietra dura Inlay work using semi-precious stones

Proterozoic 2500 to 541 million years ago

Qila Fort or a castle

Quaternary 2.588 ± 0.005 million years ago to the present

Rubble Pieces of rough or undressed stone used in building walls

Sabz Burj Green dome

Sawan and Bhadon Indian seasons connected with rains

Seljuk A member of any of the Turkish dynasties which ruled Asia Minor in the 11th to 13th centuries, successfully invading the Byzantine Empire and defending the Holy Land against the Crusaders

Serai An inn with a central courtyard for travelers in the desert regions of Asia or North Africa

Souq A street market

Squinch A system of construction whereby series of arches rest over a square chamber turning it into an octagon that enables capping with a dome roof

Stucco A fine plaster using aggregates and binding material

Takht Royal seat
Vault Self-supporting arched roof profile
Vijay-Stambh Pillar of victory
Wazir Minister
Zanjir Chain
Zenana Women's quarters

Natural Stone and World Heritage

1. Natural Stone and World Heritage:
   Salamanca (Spain) (2019)
   Dolores Pereira
   ISBN: 978-1-138-49954-6 (HB)

Milton Keynes UK
Ingram Content Group UK Ltd.
UKHW051011071024
449327UK00001B/4